AF707754

Value Creation in Management Accounting and Strategic Management

Diverse and Global Perspectives on Value Creation Set

coordinated by
Nabyla Daidj

Volume 2

Value Creation in Management Accounting and Strategic Management

An Integrated Approach

Satoshi Sugahara
Nabyla Daidj
Sumitaka Ushio

iSTE

WILEY

First published 2017 in Great Britain and the United States by ISTE Ltd and John Wiley & Sons, Inc.

Apart from any fair dealing for the purposes of research or private study, or criticism or review, as permitted under the Copyright, Designs and Patents Act 1988, this publication may only be reproduced, stored or transmitted, in any form or by any means, with the prior permission in writing of the publishers, or in the case of reprographic reproduction in accordance with the terms and licenses issued by the CLA. Enquiries concerning reproduction outside these terms should be sent to the publishers at the undermentioned address:

ISTE Ltd
27-37 St George's Road
London SW19 4EU
UK

www.iste.co.uk

John Wiley & Sons, Inc.
111 River Street
Hoboken, NJ 07030
USA

www.wiley.com

© ISTE Ltd 2017
The rights of Satoshi Sugahara, Nabyla Daidj and Sumitaka Ushio to be identified as the authors of this work have been asserted by them in accordance with the Copyright, Designs and Patents Act 1988.

Library of Congress Control Number: 2017948831

British Library Cataloguing-in-Publication Data
A CIP record for this book is available from the British Library
ISBN 978-1-84821-976-2

Contents

Introduction

This book is devoted to an analysis of linkages between management (or managerial) accounting and strategic management in order to obtain a better understanding of value creation and capture. Most existing studies and books have adopted a purely account-based approach or a strategic-oriented approach to address this issue. We have chosen to overcome this classical divide.

Several institutes and associations in the accounting and management consulting professions have highlighted these links. According to the Institute of Management Accountants (IMA):

> "Management accounting is a profession that involves partnering in management decision making, devising planning and performance management systems, and providing expertise in financial reporting and control to assist management in the formulation and implementation of an organization's strategy".

The American Institute of Certified Public Accountants (AICPA) states that management accounting as a practice extends to the following three areas:

– strategic management: advancing the role of the management accountant as a strategic partner in the organization;

– performance management: developing the practice of business decision making and managing the performance of the organization;

– risk management: contributing to frameworks and practices for identifying, measuring, managing and reporting risks to the achievement of the objectives of the organization.

This book looks at management accounting changes and at the emerging role of management accounting in strategy making. One of the main purposes of management accounting is to help an organization achieve its strategic objectives. The book appraises how closely related accounting and strategic management could create value for companies operating their activities in a dynamic and unforeseeable business environment.

What is management accounting and what strategic role does it play in an organization? Since the 1980s, many changes have occurred and companies have focused their strategy more and more on value creation. Consequently, new strategic directions have emerged, especially for managerial accounting. Management accounting and alignment with strategy can improve performance.

The topic is related to the evolution of the objectives of management accounting based on a strategic approach (external and internal diagnosis), in order to make successful long-term decisions, to ensure a sustainable competitive advantage and to create value. According to [HIL 05], management accounting is able to:

> "'add value' to a business, through the following five major goals: providing information for decision making and planning, and proactively; participating as part of the management team in the decision-making and planning processes; assisting managers in directing and controlling operational activities; motivating managers and other employees toward the organization's goal; measuring the performance of activities, subunits, managers, and other employees within the organization; assessing the organization's competitive position, and working with other managers to ensure the organization's long-run competitiveness in its industry".

This book is divided into three parts.

Part 1 discusses the various meanings of value (creation) according to several theoretical corpus including mainly economics and strategic management in Chapter 1 and opens the debate on linkages between management accounting and strategy in Chapter 2.

Part 2 describes the evolution of the conditions of value creation from different aspects. Chapter 3 explores how management accounting systems and their practices contribute to continuous value creation by encouraging organizational learning and presents a Japanese case study. Chapter 4 raises the question of how value will be created in a context of digital transformation that reshapes value chains, business models and more broadly business practices including accounting activities. The chapter identifies structural changes relating to the advent of digital technologies and several implications of digitization in the French economy.

Part 3 investigates the factors to determine the voluntary choice of accounting standard for small and medium entities (SMEs) in Japan in order to get a more comprehensive overview of value creation. Chapter 5 summarizes the SME Accounting Scheme in Japan, the theoretical foundations of the voluntary disclosure choice including a literature review. In Chapters 6 and 7, based on empirical tests, Japanese SMEs' strategic behaviors are addressed by focusing on their choice of accounting standards. The choice of accounting standards is regarded as one of the strategic activities implemented by the firms.

This book is an attempt to adopt a cross-disciplinary approach and to explore two combined approaches (strategy and accounting) to improve our knowledge of value creation in various contexts. It draws upon a number of well-defined theoretical and empirical backgrounds and methodologies. This book encourages further thought and reflection on these issues and should be pursued in the future as firms should face new challenges with the acceleration of the digital transformation:

> "A digital transformation strategy impacts a company more comprehensively than an IT strategy and addresses potential effects on interactions across company borders with clients, competitors and suppliers" [HES 16].

Part 1

The Evolution of the Concept of Value Creation in Accounting and Strategy (At a Theoretical Level)

1

Value Creation: A Polysemic Concept

1.1. Introduction

This chapter builds a bridge between mainstream economic theory and strategic management regarding the concept of value. Economics has for a long time been an important source of ideas for developing mainstream strategic management theory. More specifically, industrial economics (known also as industrial organization [IO]) has inspired numerous scholars in strategic management. IO is related to the structure of, and boundaries between, firms and markets. The IO approach is based on economic theory and deals with issues such as competition, rivalry and resource allocation. IO is a branch of economics that emphasizes interdependence that characterizes the firms' decisions in their markets.

"Traditional" insights from economics are closely related with the use of "common" concepts such as value. This chapter is dedicated to a discussion of the concept value and its extension in the field of strategy. The design of this chapter is as follows. We begin by explaining the basics of the notion of value from the economic perspective before developing its main meanings in various contexts.

1.2. The economic concept of value

1.2.1. *Back to the basics*

The history of economic thought has been concerned with economic *value*. Economic goods either have use value or exchange value. The use value of a commodity is considered as the direct utility that one receives from its consumption.

> "It has reference to the needs which the properties of a commodity as a physical artifact can be employed to cater to" [GID 71]

The exchange value is the quantitative aspect of value, i.e. the quantified worth of one good or service expressed in terms of the worth of another. "Use value" and "exchange value" can be illustrated by the famous "water-diamond" paradox, i.e. that diamonds are naturally more valuable than water *not* because diamonds are more expensive to produce, but rather because diamonds are more scarce than water.

It is worth noting that this debate remains a key issue in the 2000s and 2010s in the field of marketing and strategic management (see section1.2.2). As mentioned by [KRA 11]:

> "A useful contribution to this debate is Bowman & Ambrosini's (2000) distinction between 'perceived use value' (the subjective value perceived by customers) and 'exchange value' (the bargained price that is paid). The distinction is useful because it emphasizes that the use value of a product/service is a perceived value and that this perceived value may differ from the price that is paid".

Ricardo [RIC 51] and Smith [SMI 81] were the first authors to make a distinction between use-value and exchange-value, focusing their attention on the latter. Then, Marx adopted a Hegelian perspective (based on labor theory) and considered use-value and exchange-value as inseparable dialectical aspects of "the commodity". The neoclassical school [JEV 71, MEN 71, WAL 74] highlighted that value should be determined for each commodity taken separately. The exchange value is considered as a function of use value of the utility of the given commodity. Then, the transformation of value into prices must be done. Value reappears in 20th century economics, as in the expression of "value-adding".

Economists have also developed another concept named "economic rent". Ricardo defined rent as "that portion of the produce of the earth which is paid to the landlord for the use of the original and indestructible powers of the soil" Marshall added, "the income derived from the ownership of land and other free gifts of nature is commonly called rent".

Modern economists use the word rent as an "economic surplus or transfer earnings that means the earning of a factor of production in excess of the minimum amount necessary to keep it in its present use". Rents can be generated from resources, innovation advances and/or specific market structure (monopoly rents allowing the company to generate significant and exceptional returns).

Schumpeter [SCH 34] is a key figure on technological innovation and is considered "the prophet of innovation" (an expression used by Thomas McKraw in [MCK 10]). He argued that economic development is driven by innovation through a dynamic process in which new technologies replace the old, a process he named "creative destruction". In Schumpeter's view, "radical" innovations create major disruptive changes, whereas "incremental" innovations continuously advance the process of change. Schumpeter [SCH 34] proposed a list of five types of innovations: new products, new methods of production, new sources of supply, the exploitation of new markets and significant changes in workplace organization and management. Schumpeterian competition drives innovation.

From the Schumpeterian perspective:

> "economic logic prevails over the technological" [SCH 34]
>
> "Costs as an expression of the value of other potential employments of means of production constitute the liability items in the social balance sheet. This is the deepest significance of the cost phenomenon" [SCH 34].

1.2.2. *The concept of value explained by IO scholars*

As it has been mentioned in section 1.1, IO takes into consideration several markets (monopoly, duopoly and oligopoly), product differentiation, incomplete information and various strategic variables (price, advertising, R&D, capacity) in order to have a better understanding of strategic interdependence. IO models, to a large extent, adopt an external perspective to explain how the external environment (government decisions, industry) influences firms' strategic actions and interactions between them. There are different approaches to industrial organization: theories of the firm (transaction cost theory, agency theory and economics of property rights), game theory, etc.

1.2.2.1. *The notion of value in game theory*

Game theory is a branch of applied mathematics in relation to economics that considers strategic interactions between agents in a context of uncertainty. It is the mathematical approach of IO. Game theory is mainly used for analyzing market structures or the behavior of various players (firms, States, institutions, regulatory authorities, etc.) and formalizing the rivalry that characterizes their relationships or the cooperation processes they want to build. It postulates that agents choose strategies to maximize their payoffs, given the strategies of other agents. Game theory, based on a process of iterations, has been applied to cooperation issue.

Games can be *non-cooperative* or *cooperative*. In non-cooperative games, players move according to an individualistic behavior, without taking into account the general interest, i.e. they exclusively pursue their own interests.

Cooperative games describe specific situations in which players jointly seek a collectively satisfying solution. In this framework, communication among players is possible. In this field, games are often based on the "cake-sharing" problem, i.e. the allocation of the aggregated outcome that results from a common cooperative action: "core" notion, Nash bargaining solution [NAS 50]. It is about determining an allocation that gives each player (or initial coalition of players) at least the payoff that he might individually obtain through an independent action. Nevertheless, as shown by the classical *prisoner's dilemma*, players' individualistic strategies make it difficult to implement the cooperative solution. The game's outcome does not always represent the collectively optimal (Pareto optimal) solution. Hence, when that the agreement is closed and each player is sure that the others respect it, he may attempt to unilaterally deviate. This deviation (or treason) strategy enables him to obtain a better payoff than that of cooperation. This phenomenon has been treated with scepticism by game theorists, who wonder whether it is possible to stabilize an agreement when players are free to act.

The solution to implementing the "cooperative" situation, i.e. the situation that improves each player's outcome with respect to the non-cooperative situation, can be to modify the game by introducing a third party that punishes deviations or to consider an infinitely repeated game that might enable players to "self-punish" deviations that are observed at a given game's stage (see, for example, [FRI 71, LAM 98, ABR 88, FUD 86]).

In this vein, the theory of endogenous coalition formation [HAR 83, BAL 00, BLO 95, BLO 96, RAY 97, RAY 99] argues that a stable cooperation/ coordination is that to which players spontaneously adhere without constraints or irreversible commitments. Hence, only *self-enforcing* cooperation/ coordination is stable, i.e. it is not threatened by players' unilateral deviations, because it results from players' voluntary adhesions.

The non-cooperative approach to coalition formation thus puts forward the idea that coalitions result from players' decisions rather than from the negotiation of contractual agreements. Hence, each player decides whether to adhere to a cooperative (or to a coordinated project), or in more difficult cases, he chooses the coalition to be part of, without being required to respect any agreement. The decision to "subscribe", or "cooperate", thus represents a strategic variable like any other chosen within the framework of a specific non-cooperative game, the related strategic space being taken into account. Coalitions thus emerge as the outcome of a non-cooperative game through voluntary players' adhesions. A coordination that pertains to the adhesion variable does not imply players' commitment, which is rather related to the cooperative approach to coalition formation. More generally, we can state that coalition formation is endogenous.

Box 1.1. *Game theory and strategic management (adapted from [DAI 10])*

Grant [GRA 91] considers that game theory has several valuable contributions to make to strategic management:

> "1- it permits the framing of strategic decisions. Apart from any theoretical value of the theory of games, game theory provides a structure, a set of concepts, and a terminology that allows us to describe a competitive situation in terms of identity of the players, specification of each player's options, specification of the payoffs from every combination of options, the sequencing of decision using game trees. 2- It can predict the outcome of competitive situations and permits the selection of optimal strategic choices".

In addition, game theory uses the Shapley value notion. The Shapley value was proposed by Shapley in his 1953 PhD dissertation and then developed further. "The value of a game [that] depends only on its abstract properties". The Shapley value is a classic cooperative solution concept.

> "It is a solution that prescribes a single payoff for each player, which is the average of all marginal contributions of that player to each coalition he or she is a member of. It is usually viewed as a good normative answer to the question posed in cooperative game theory. That is, those who contribute more to the groups that include them should be paid more" [SER 07].

1.2.2.2. *The theories of the firm and the concept of value*

The theory of firm (ToF) has several branches: transaction cost approach; "rent-seeking" theory, agency theory and, "property-rights" theory. Several concepts have been developed within this framework: transaction costs; firms *versus* markets; different forms of contracts issue; the representation of a firm as a nexus of incomplete contracts. These theories aim to explain internal firm organization and market in defining transaction costs, property rights and enforcing contracts. Transaction cost economics [WIL 75, WIL 85] and the modern property rights models [GRO 86, HAR 90] have also developed the concept of economic value. The firm's property rights, third-party enforcement and self-enforcing agreements protect the process of creating resources and its related value.

Coase [COA 37] was one of the first scholars who posed the following questions: why firms exist? What precisely a firm is? And which transactions are more efficiently conducted in a firm than in a market? He [COA 37] considered that the firm and the market should be viewed as alternative ways of organizing transactions. According to him, resources can be allocated in two ways: either via the market or via the firm. The originality of this analysis lies in that the author acknowledges the existence of operating costs in the market. The system that relies on market-specific prices leads to transaction costs, which justifying the existence of firms, who are then responsible for minimizing those same costs:

> "Outside the firm, price movements direct production, which is co-ordinated through a series of exchange transactions on the market. Within a firm, these market transactions are eliminated and in place of the complicated market structure with exchange transactions is substituted the entrepreneur-co-ordinator, who directs production. It is clear that these are alternative methods of co-ordinating production" [COA 37].

In addition, the first formal statement of the Coase theorem did not come until 1966, when [STI 66] offered that:

> "The Coase theorem ... asserts that under perfect competition private and social costs will be equal".

The Coase theorem describes the economic efficiency of an economic allocation or outcome in the presence of externalities, perfect information and zero transaction costs.

Then, [WIL 81] continued these studies and elaborated a transaction cost theory based on the idea that the existence of the firms can be explained in terms of the cost of carrying out transactions across markets. The firm is assumed to create value by reducing costs in comparison to the market.

Agency theory [JEN 76] explains that the firm can be defined as a nexus of contracts, and therefore a legal fiction and describes situations and relationships in which one party (the principal) delegates work to another (the agent). The main objective of agency theory is to explain how explicit or implicit contracts can be drawn up between the two parties to take account of shirking, opportunism, bounded rationality and imperfect and incomplete information to monitor agent behavior and to propose an optimal incentive structure. Agency theory has been applied to a variety of strategic management topics such as corporate strategy, corporate governance, etc. In modern corporations characterized by separation of ownership and control, the interests of shareholders and managers may diverge. In this context, managers will seek to maximize their own interests at the expense of shareholders. According to agency literature on corporate governance, the board of directors is a control instrument to protect shareholders' interests in the value distribution process.

Alchian [ALC 65] and Alchian and Demsetz [ALC 73] analyzed the structure of property rights in a society and the consequences for social interaction from a specific structure of property rights. Property rights are defined as:

> "[t]he rights of individuals to the use of resources… supported by the force of etiquette, social custom, ostracism, and formal legally enacted laws supported by the states' power of violence or punishment" [ALC 65].

All economic activities including trade and production are the exchange of bundles of property rights [FUR 72]. Property rights theory can improve understanding of analysis of both firms and governance practices.

> "Conjecture that with advances in coordination enabled by information technology, intellectual property rights research will be a growth industry for at least the next decade in the discipline of strategic management" [KIM 05].

In addition, Kim and Mahoney [KIM 10] have considered that resource-based theory or resource-based view (RBV) and property rights theory are complementary providing a complete view of economic value creation and the distribution of economic value.

Kraaijenbrink and Spender [KRA 11] have suggested a synthetic and analytical grid including several theoretical approaches presented briefly previously in order to get a better understanding of value. They have analyzed value creation from two dimensions, the supply and the demand sides (Table 1.1). According to the authors,

> "The first dimension concerns the supply of value. On this dimension the ToFs differ in their assumptions on whether value is an inherent property of assets that needs to be extracted, or that value is actively created by firms. The resource-based view is based on the first assumption. It presumes that value is an inherent property of resources and that this value needs to be discovered. On the other hand, the Austrian economic view on the firm presumes that the value of assets lies in their subjectively recognized attributes (…).The second dimension on which we can compare how the ToFs deal with firms' value creation concerns the demand for value. On the one hand, and adopting a Kirznerian (1973) view, there are ToFs which assume the demand for value can be anticipated and predicted. The assumption is that, somewhere in the market, there is an unfulfilled need, which needs to be discovered and then fulfilled. On the other hand, adopting a more Schumpeterian (1934) view, there are ToFs that assume the demand for value does not pre-exist, but that it is created".

<table>
<tr><th>Demand for value
Supply of value</th><th>Predicted/discovered</th><th>Created</th></tr>
<tr><td rowspan="2">Extracted/discovered</td><td>Bureaucratic theory
Team production
Transaction cost economics</td><td></td></tr>
<tr><td colspan="2">Behavioral theory
Entrepreneurial theories of firms (ToFs)</td></tr>
<tr><td>Created</td><td>Value chain</td><td>Austrian economics</td></tr>
</table>

Table 1.1. *Comparison of extant theories of the firm ([KRA 11])*

1.3. The value concept in strategic management: the evolution of strategic thought of Michael Porter

Since the 1950s, the issue of value has been addressed by several researchers in strategy. One of the currently dominant views of strategic management – RBV of firms – is based on the concept of economic rent. From a resource-based perspective [BAR 91, PET 93], firms are able to create economic value (i.e., generate economic rents) and sustainable advantage by developing resources that are valuable, rare, difficult to imitate and non-substitutable.

We propose in this section to focus our attention on the evolution of strategic thinking of one of the most influential scholars, Michael Porter, regarding the notion of value. Porter [POR 85] analyzed how the value is created and captured by a company. He therefore introduced the concept of the value chain as the basic tool for examining the activities a company performs and their interactions in order to identify the source of sustainable competitive advantage. Porter also derives the concept of "profit margin", which is the difference between total value (created and captured) and the collective cost of performing the value activities. A company aims to maximize the value offered to customers.

For decades, Porter has elaborated strategic concepts and tools established various frameworks (five forces, value chain, etc.), which have been widely disseminated and commented on by faculty and other front line staff in academic centers. Consequently, the purpose of this section is not to provide a synthesis of critical reviews, but rather, to show the evolution of Michael Porter's thought based on several academic papers and main books. This evolution can be subdivided into several phases.

1.3.1. *The first stage: the economical background of Porter*

To illustrate the close links between economics and strategy described in earlier sections of this chapter, it is interesting to note that Porter's first works in the 1970s were inspired in large part by industrial economists' research in the field of industrial organization. Box 1.2 presents the most emblematic documents of the economic issues addressed by Porter between 1974 and 1982.

1974 – M.E. Porter, "Consumer Behavior, Retailer Power and Performance in Consumer Goods Industries", *Review of Economics and Statistics*, vol. 56, no. 4.

1975 – M.E. Porter, Note on the Structural Analysis of Industries. Harvard Business School, Note 376-054, September.

1976 – R. E. Caves, M.E. Porter, "Barriers to Exit", in: Joe Staten Bain, Robert T. Masson and P. David Qualles (eds.), *Essays on Industrial Organization in Honor of Joe S. Bain*, Ballinger Press, Cambridge, MA.

1976 – M.E. Porter, "Please Note Location of Nearest Exit: Exit Barriers and Planning", *California Management Review*, vol. 19, no. 2.

1977 – R. E. Caves, M.E. Porter, "From Entry Barriers to Mobility Barriers: Conjectural decisions and contrived deterrence to new competition", *The Quarterly Journal of Economics*, vol. 91, no. 2.

1979 – M.E. Porter, "How Competitive Forces Shape Strategy", *Harvard Business Review*, vol. 57, no. 2.

1979 – M.E. Porter, "The structure within industries and companies' performance", *Review of Economics and Statistics,* vol. 61, no. 2.

1980 – M.E. Porter, *Competitive Strategy: Techniques for Analyzing Industries and Competitors*, Free Press, New York, NY.

1981 – M.E. Porter, "The Contributions of Industrial Organization to Strategic Management: A Promise Beginning to Be Realized", *The Academy of Management Review*, vol. 6, no. 4.

1982 – M.E. Porter, "Price Wars, More Liberal Credit and Other Competitive Maneuvers", *Boardroom Reports*, vol. 11, no. 1.

1982 – M.E. Porter, "Industrial Organization and the Evolution of Concepts for Strategic Planning", in: T. H. Naylor (Ed.), *Corporate Strategy*, Free Press, New York, NY.

Box 1.2. *Main publications' titles in the 1970s*

1.3.2. *The second phase: a shift toward strategic management (1980–2000)*

Since the end of the 1970s and the beginning of the 1980s, Michael Porter has developed several theoretical concepts and operational tools that could help to provide a better understanding of company competitive strategy and sustainable competitive advantage. Since this decade, these concepts have become well known and widely shared. Two of them are presented below. The first one (the five forces' framework) is closely linked with the external diagnosis and the second one (company value chain) aims to analyze the internal level.

1.3.2.1. *Toward a sustainable competitive advantage*

Firms compete in international markets. How do firms create and sustain competitive advantage? At the heart of positioning is competitive advantage. In the long run, firms succeed relative to their competitors. There are two basic types of competitive advantage:

– lower cost: is the ability of a firm to design, produce and market a comparable product more efficiently than its competitors. At prices at or near competitors, lower cost translates into superior returns;

– differentiation: is the ability to provide unique and superior value to the buyer in terms of product quality, special features or after-sale service. Differentiation allows a firm to command a premium price, which leads to superior profitability provided costs are comparable to those of competitors.

Porter [POR 80] argued that a business can develop a sustainable competitive advantage based on cost, differentiation or both.

> "Competitive advantage cannot be understood by looking at a firm as a whole. It stems from the many discrete activities a firm performs in designing, producing, marketing, delivering and supporting its product. Each of these activities can contribute to a firm's relative cost position and create a basis for differentiation" [POR 85].

Porter identified several drivers of uniqueness: policy and decision; linkages among activities; interrelationships among business units; integration; scale; learning; timing of market entry; geographic location; institutional factors (regulation, union activity, taxes, etc.). Porter also identified 10 cost drivers related to value chain activities: economies of scale; learning; capacity utilization; linkages among activities; interrelationships among business units; degree of vertical integration; timing of market entry; firm's policy of cost; timing of market entry; firm's policy of cost or differentiation; geographic location and institutional factors (regulation, union activity, taxes, etc.). A cost advantage also can be pursued by reconfiguring the value chain. Reconfiguration means structural changes such as a new production process, new distribution channels, etc.

However, since the beginning of the 2000s, the notion of sustainable competitive advantage has been increasingly questioned [CAR 14] by several authors. In their analysis of business ecosystems, Iansiti and Levien [IAN 04] highlighted the fragile nature of competitive advantage "in situations of significant technological and market upheaval" (p. 9). Indeed, throughout the 1990s and 2000s, questions were being increasingly asked about how sustainable competitive advantage really is when firms are faced with industrial upheavals, technological advances and an increasingly uncertain environment where the competitive dynamics were constantly being redefined.

In 1994, D'Aveni introduced the concept of hypercompetition. He explains that competitive advantage is, by definition, destined to disappear in such a context and that it is futile to attempt to defend a sustainable competitive advantage. The only sustainable position is that of movement. Thus, D'Aveni *et al.* [DAV 10] propose "the age of temporary advantage" as an alternative concept. A competitive firm should constantly be able to reposition itself in terms of its value proposition, its know-how, its financial capacity in light of the changing entry barriers and time frames of evolving competitive dynamics. Hypercompetitivity presupposes permanent transformation of competitive advantages in order to create and capture value. This idea is shared by [MCG 13] who also called into question the relevance of sustainability in today's fast moving and hypercompetitive marketplaces.

1.3.2.2. *The company value chain*

The well-known value chain framework is a model developed by Porter [POR 80]. The generic model is related to the activities implemented by the company. More recently, the value chain representation has been extended to an entire industry. These two dimensions – internal and external – can be described in a value chain. The use of value chain analysis facilitates the strategic positioning of an organization. The value chain provides a useful tool to operationalize and implement the generic strategy.

1.3.2.2.1. The generic model

To understand the activities through which firms can create a competitive advantage and generate shareholder value, the business system can be divided into several value-generating activities referred to as the value chain. This concept was created by M.E. Porter in his book "Competitive Advantage: creating and sustaining superior performance". In a later article about the Internet, he summarized the value chain as follows:

> "When a company competes in any industry, it performs a number of discrete but interconnected value-creating activities, such as operating a sales force, fabricating a component, or delivering products, and these activities have points of connection with the activities of suppliers, channels, and customers. The value chain is a framework for identifying all these activities and analyzing how they affect both a company's costs and the value delivered to buyers" [POR 01].

The generic model represents all the internal activities a firm develops to produce goods and services. It helps to analyze support activities that add value to the final product indirectly and primary activities ("specific activities" or "core activities") through which firms can create value and competitive advantage (based on costs and/or differentiation). It is used to analyze which activities are the most valuable. Value chain analysis can help a company determine which type of competitive advantage to pursue, and how to pursue it.

The value chain is a useful analysis for defining a firm's core competencies and the activities in which it can pursue a competitive advantage (as mentioned previously) as follows: cost advantage by better understanding costs and squeezing them out of the value-added activities, and differentiation by focusing on those activities associated with core competencies and capabilities in order to perform them better than competitors. As resources and competencies are rare, valuable, specialized, hard to access, difficult to imitate and non-substitutable, they often constitute strategic assets from which the company's competitive advantage stems over its rivals. The RBV is very useful to complete the value chain approach.

Porter used the word "margin" for the difference between the total value and the cost of performing the value activities. This margin depends on firms' ability to make the linkages between all activities in the value chain. Here, value is referred to as the price that the customer is willing to pay for a certain product or service delivered by the organization [MAC 00] and to pay more than the sum of the costs of all activities in the value chain. Other scholars have used the word "added value" instead of margin [LYN 03].

1.3.2.2.2. Limitations of value chain analysis

The limitations of value chain analysis concern its implementation and interpretation as well:

– product-oriented analysis: one of the limitations of the value chain model is that it describes an industrial firm that purchases raw materials and transforms them into physical products. Its applicability in the context of service organizations seems more difficult;

– identification of value and costs: finding the costs, revenues and assets for each value chain activity is difficult. Several authors consider that this work is done through trial-and-error and experimentation methods. In addition, this activity requires regularly updating data. The value chain analysis must be repeated to take into account changes in cost structures, technology, market prices and capital investments from one period to the next;

– partnerships/networks: firms use different types of partnerships (collaboration, strategic alliances and outsourcing) and networks (business ecosystems) to gain advantages in cost, quality, time, flexibility, delivery and technology. These relational practices cause major changes in organizations and their value chains but these are not included in the value chain analysis;

– international dimension: many firms nowadays conduct international operations all around the world (products exports, production abroad spread over several geographical locations, licensing agreements) and this dimension is not mentioned in value chains;

– information: a value chain requires in-depth analysis. It is actually difficult for a company to collate relevant internal information in order to represent and comment on its own value chain (company value chain) and it is also a complicated task to collect external data and information on its main rivals (industry value chain).

1.3.3. *The third phase: the role of geographic location (the 2000s), clusters and nations*

In the 2000s, Porter broadened his focus to include other subjects related mainly to "location" of activities. Several papers published during this period (Box 1.3) attest to this evolution. The development of competitive advantage is closely related to the location. This approach is logical as strategic management refers to the entire scope of strategic decision-making activity in an organization closely linked with its external environment. In its strategic decision making, a firm has to take into consideration three levels of analysis: country, market (sector) and corporate level [DAI 15].

1990 – M.E. Porter, *The Competitive Advantage of Nations*, Free Press, New York, NY.

1994 – M.E. Porter, "The Role of Location in Competition", *Journal of the Economics of Business*, vol. 1, no. 1.

1997 – A.M. McGahan, M.E. Porter, "How Much Does Industry Matter, Really?", *Strategic Management Journal*, vol. 18.

1998 – M.E. Porter, Ö. Sölwell, "The role of geography in the process of innovation and the sustainable competitive advantage of firms", in: Alfred D. Chandler, P. Hagström and Ö. Sölwell, (eds.), *The Dynamic Firm—The Role of Technology, Strategy Organization, and Regions*. Oxford University Press, Oxford.

1998 – M.E. Porter, "Clusters and the new economics of competition", *Harvard Business Review*, vol. 76, no. 6.

1999 – M.E. Porter, "Clusters and the New Economy", in Charles Edquist and Maureen McKelvey (eds.), *Systems of Innovation: Growth, Competitiveness and Employment*, Edward Elgar Publishers, Ltd. Cheltenham.

2000 – M.E. Porter, "Location, competition and economic development: local clusters in the global economy", *Economic Development Quarterly*, vol. 14, no. 1.

Box 1.3. *The role of location*

Porter's interest in this notion of location emerged when he started working on competitive advantage of nations at the end of the 1990s. Up until this time, he had mainly focused on competitive advantage of companies (see above) but his research progressively integrated multiple and embedded levels of analysis (company, country) consistent with market and industry changes.

In addition, as complex relationships between firms are an increasingly prevalent and important trend in business practice, he has chosen to study one specific type of interorganizational networks called "clusters", relational organizational form that involve systems of various interconnected players. Most clusters are located in regions or territories.

1.3.4. *The fourth phase: from the end of the 2000s to the beginning of the 2010s*

This period has been characterized by a new direction adopted by Porter in his strategic reflections. In his most recent publications, the author addressed two main issues related to the link between competitive advantage and corporate social responsibility and to the concept of value.

Regarding the second topic, Porter discussed the notion of shared value in order to broaden the understanding of value. According to [POR 11]:

> "The purpose of the corporation must be redefined as creating shared value, not just profit per se. This will drive the next wave of innovation and productivity growth in the global economy. (…). Companies must take the lead in bringing business and society back together. (…). The solution lies in the principle of shared value, which involves creating economic value in a way that also creates value for society by addressing its needs and challenges. Businesses must reconnect company success with social progress. Shared value is not social responsibility, philanthropy, or even sustainability, but a new way to achieve economic success. It is not on the margin of what companies do but at the center. We believe that it can give rise to the next major transformation of business thinking".

Even if there have been few empirical analyses of the use of shared value so far, this concept could be considered as a step forward in the evolution of strategic thought as it refers to a broader vision of business. Both economic and social/societal aspects are factors to be considered by firms in order to achieve competitive advantage.

> "The concept of shared value (...) recognizes that societal needs, not just conventional economic needs, define markets. It also recognizes that social harms or weaknesses frequently create internal costs for firms – such as wasted energy or raw materials, costly accidents, and the need for remedial training to compensate for inadequacies in education. And addressing societal harms and constraints does not necessarily raise costs for firms, because they can innovate through using new technologies, operating methods, and management approaches – and as a result, increase their productivity and expand their markets. Shared value, then, is not about personal values. Nor is it about "sharing" the value already created by firms – a redistribution approach. Instead, it is about expanding the total pool of economic and social value" [KRA 11].

1.4. From value creation to cocreation of value

Practices such as co-conception, codesign, cocreation and coproduction all developed in the early 2000s have been analyzed by many authors in the field of strategic management and marketing as well. They are very interesting as they challenge the "traditional" terms and conditions of value creation.

These "new" approaches have been also adopted by many companies from various sectors. In varying degrees, they all imply a collaborative, collective and interdisciplinary dimension. Instead of relying upon a "technology push" philosophy, which prevailed for a long time, they confide in a "market pull" outlook, which seeks to integrate market needs with the expectations of final users.

Vargo and Lusch [VAR 04, VAR 08] consider that there are alternative logics for understanding markets. Consequently, they view service rather than goods as the focus of economic and social exchange. Service (singular) is a process distinct from "services". Services are particular types of goods characterized by: intangibility, heterogeneity (non-standardization), inseparability (of production and consumption) and perishability.

> "The most critical distinction between G-D logic and S-D logic is found in the conceptualization of service. In S-D logic, service is defined as the application of competences

(knowledge and skills) for the benefit of another party. The use of the singular 'service' as opposed to the plural 'services', as traditionally employed in G-D logic, is intentional and non-trivial. It represents a shift from thinking about value in terms of operand resources—usually tangible, static resources that require some action to make them valuable – to operant resources – usually intangible, dynamic resources that are capable of creating value. That is, whereas G-D logic sees services as (somewhat inferior to goods) units of output, S-D logic sees service as a process – doing something for another party. The locus of value creation, then, moves from the 'producer' to a collaborative process of co-creation between parties" [VAR 08].

The concept of co-creation of value is highlighted by [VAR 08]: value is only created with and determined by the user. The enterprise can only make value propositions, collaborating with customers and partners to create and sustain value.

GDL concepts	Transitional concepts	SDL concepts
Goods	Services	Service
Products	Offerings	Experiences
Feature/attribute	Benefit	Solution
Value added	Coproduction	Cocreation of value
Profit maximization	Financial engineering	Financial feedback/learning
Price	Value delivery	Value proposition
Equilibrium systems	Dynamic systems	Complex adaptive systems
Supply chain	Value chain	Value-creation network
Promotion	Integrated marketing communications	Dialogue
To market	Market to	Market with
Product orientation	Market orientation	Service orientation Interacting

Table 1.2. *Goods-dominant logic, service-dominant Logic and transitional concepts ([LUS 06])*

1.5. Conclusion

The concept of value, rich in meaning, has been adopted based on various definitions by economists, lawyers, accountants and financial analysts. Increasingly, the notion has been has become more inclusive to encompass both economic and broad societal implications. The development of shared value is an outstanding example of this evolution.

2

Strategy and Management Accounting: Theoretical Background[1]

2.1. Defining "strategy"

What is strategy? This question has been asked for many years. The origin of the concept of strategy goes back to Sun Tzu, a Chinese military strategist who wrote in the 5th Century BC. Even today, discussions about strategy are diverse and the definition of strategy varies. To make matters worse, the word "strategy" sometimes refers to fundamentally different concepts of dimensions. The aim of this chapter is to overview the concepts and definitions of strategy as well as the research regarding the role of accounting in terms of strategy. According to Mintzberg *et al.* [MIN 98], there are five definitions of strategy, with fundamental concepts behind them. This chapter begins with a description of these definitions.

The first definition states that "strategy is a plan". Strategy is an aggregation of future actions that are expected to achieve organizational objectives. This approach requires an analysis of the current situation and a prediction of the future in order to clarify the actions that are needed to achieve a desired result. Such a strategic approach is also referred to as "intended strategy". The alternative name clarifies the features of this line of strategy [CHA 05].

The second definition states that "strategy is a pattern". In this regard, strategy is a collection of actions taken consistently in the past. Such a

1 This work was supported by JSPS KAKENHI, grant number 16K17216.

strategic approach may unintentionally form and/or differ from prior planning. An organization's members focus their capabilities on the tasks and realities that they face in their daily activities, thereby accumulating actions that they consider desirable from time to time. The pattern of actions and behaviors generated in this way is the strategy for the organization. What is important in this definition is the creation of a mechanism that encourages the capabilities of an organization's members rather than the preparation of clear prescriptions (i.e., strategy as a plan) in advance. Davila [DAV 05] refers to this approach as "emergent strategy" in contrast to "intended strategy".

The third definition states that "strategy is a position". This definition is the idea that the way in which an organization's products and services are positioned in the market is the key to attaining competitive advantage. Such an approach argues that the profitability of a business is determined by two factors: the potential and long-term profitability of the industry itself and the relative profitability of the organization within that industry.

The fourth definition states that "strategy is a perspective". This definition refers to the basic idea and philosophy of a business or organization. Thus, strategy is a concept that an organization pursues in the long term. Such an approach does not conflict with the third idea (i.e., "strategy is a position") since these two represent different notions of different dimensions of the same word, "strategy" (which, however, makes the situation even more confusing). For example, a hamburger restaurant chain may launch new muffins in a less competitive breakfast market (i.e., the restaurant chain may take a new position) in terms of "providing reasonable and delicious items quickly". One point to keep in mind is that developing a new market and sustaining a concept are referred to by the same word, "strategy".

The fifth definition states that "strategy is a ploy". This definition refers to a maneuver to cheat competitors. Although there may be a purpose behind the maneuver in most cases, deceiving the target is sometimes referred to as a "strategy" or a "strategic" move.

Thus, the term "strategy" occasionally refers to concepts with different dimensions. This variety of definitions requires writers and readers to pay careful attention in order to avoid confusion. The only aspect in common is that, as Mintzberg *et al.* [MIN 98] say, any concept of "strategy" refers to "something [that relates] to [the] common overall welfare of the organization". Based on the foregoing, section 2.2 provides further discussion about the ideas behind strategy.

2.2. The "prescriptive view" and "postscriptive view" of strategy

2.2.1. *The prescriptive view of strategy*

The concept of strategy as a plan or position may be familiar to many readers. In this regard, the strategic process starts with a careful analysis of the business environment. One of the most famous methods, which was proposed by Albert S. Humphrey in the 1960s, is a strategic planning tool used to evaluate the strengths, weaknesses, opportunities and threats (SWOT) of/for an organization. SWOT analysis is designed to be used in the preliminary stages of strategic management planning in order to identify the internal and external factors that are favorable and unfavorable for achieving organizational objectives [ARS 08].

Porter [POR 80, POR 85] suggests a framework to analyze an industry's structure from the perspectives of five forces: the threat of new entrants, the bargaining power of buyers, competitive rivalry, the bargaining power of suppliers and the threat of substitutes. The purpose of this analysis is to understand the nature of rivalry and profitability within an industry. Based on the analysis, Porter proposes three types of positioning to shape the competitive forces for an organization: cost leadership, differentiation and focus. In this regard, strategy is the creation of unique and valuable positioning.

Other methods that analyze the business environment, such as customer, competitor and company (3C) analysis and political, economic, social and technological (PEST) analysis, are also well known. Moreover, product portfolio management (PPM), proposed by the Boston Consulting Group in the 1970s, provides insights into how to optimize the business portfolio and resource allocation within an organization by using the categorizations of "star", "cash cow", "question mark/problem child" and "dogs".

The methods outlined above are based on an "outside-in" perspective in the sense that they examine the external environment in order to identify potential threats and opportunities for an organization. The "inside-out" perspective identifies competitive advantages derived from an organization's internal strengths [DEW 99, CHE 05].

Barney [BAR 91, BAR 95] employs the resource-based view to seek sources of competitiveness in organizations' capabilities. These sources include tangible assets, production facilities, brand names, patents and other technical and intellectual capabilities. The author provides a framework called VRIO (value, rareness, imitability and organization) to evaluate how much each source contributes to an organization's strategic planning. Based on these criteria, highly evaluated internal resources enhance the organization's sustainable competitive advantage.

A common feature among these concepts is that the reasonable actions that an organization should take are to be derived through in-depth, advance analysis of business environments and/or internal capabilities. This chapter refers to such an approach as the "prescriptive view of strategy". This strategic view is clear and (seemingly) logical. Thus, it is used quite often both in practice and for academic research. However, it is also criticized, mainly for the following two reasons.

The first is the limitation of prediction. No matter how much time and effort are undertaken in advance for an analysis, such as a SWOT, five forces or 3C analysis, it is difficult to forecast the future comprehensively. Uncertainty about prospects is a limitation on reasonable planning and proposed positioning. The second criticism is the separation of the formulation and implementation of a strategy. Even if a reasonable strategy is formulated at the management level in advance, it must be transmitted to and understood by the shop floor correctly. Even though the strategy may be defined and divided into detailed actions at shop-floor level so that employees do not need to understand the overall picture, the strategy requires enormous time and effort and may result in decreasing the prompt and flexible action that is required for adapting to changes in preconditions.

2.2.2. *The postscriptive view of strategy*

However, another strategic view emphasizes the process of trial and error in an organization. Mintzberg *et al.* [MIN 98] refer to this view as the "learning school". In this regard, strategy is a pattern of actions taken in the past within an organization. For example, Quinn [QUI 80, p. 145] contends as follows:

> By the time the strategy begins to crystallize in focus, pieces of it are already being implemented … Constantly integrating the simultaneous incremental processes of strategy formulation and

implementation is the central art of effective strategic management.

Because the strategy is created through day-to-day activities throughout an organization, its formulation and implementation are closely intertwined. More precisely, strategy is not something to be formulated by specific people or groups; instead, it is something that forms or emerges unnoticed within an organization.

What is important in this view is to recognize the collection of excellent behaviors that develop in an organization as soon as they emerge, rather than spending a great deal of time on prospects and analysis in advance, and to stretch the pattern toward the future. In contrast to the prescriptive view, this chapter refers to these concepts as the "postscriptive view".

Nonaka and Takeuchi [NON 95] suggest a framework of knowledge creation and organizational learning processes called the socialization, externalization, combination and internalization (SECI) model. They emphasize the conversion of tacit knowledge to explicit knowledge. Tacit knowledge is personal and context specific; thus, it is hard to formalize and communicate [POL 66]. According to Nonaka and Takeuchi [NON 95], tacit knowledge should be transferred and shared with others via coworking experiences and on-the-job training ("socialization"). It is also necessary to crystallize tacit knowledge by quantification and verbalization in order to convert it into explicit knowledge ("externalization"). "Externalization" makes it easier to transmit knowledge to other employees and combine it with other knowledge ("combination"). By utilizing transferred explicit knowledge, such knowledge becomes part of other employees' knowledge, which results in the creation of a new tacit knowledge within an organization ("internalization").

In order to facilitate trial and error, and encourage ingenuity at the shop-floor level, an organization's shared values or credos (or "perspective" according to Mintzberg *et al.* [MIN 98]) have an important role. Nonaka and Toyama [NON 07] argue that the "common good" must be shared and well understood in an organization in order to create wisdom or "phronesis" at the local level. For example, in an automobile-producing company, employees may work hard in order to "produce a good car". However, the concept of "a good car" probably varies for each employee. Some people may imagine that a good car has the operational ability and acceleration of a racing car in order to make the customers' lives enjoyable or is a car that demonstrates fuel economy with hybrid technology for a sustainable society. Other employees may think that a good car is a vehicle that arrives at a destination by using an automatic driving function or is a car that can be easily mass produced with low costs so that it can be diffused in developing countries. The directions of employees' endeavors are significantly affected by the perspective for which an organization aims.

As with the prescriptive view, there are criticisms of such a view of strategy. The first concerns inefficiency. As well as ensuring organizational learning through trial and error, it is important to conduct daily tasks efficiently. There is even a possibility that nothing valuable emerges by leaving matters to the shop floor. In that case, it is much better to execute activities and routine work arranged by senior managers obediently without in-depth thought at the local level as long as they are based on accurate forecasts to some extent. The second criticism is what Johnson [JOH 87] calls "strategic drift". Relying on spontaneous actions and responses at the local level may result in a lack of consistency within an organization. In particular, long-term decisions, such as mergers and acquisitions (M&A) and significant investments, cannot be made through creative ingenuity or trial and error at the shop-floor level alone.

The disadvantages of prescriptivism and postscriptivism are the reverse of their merits. In reality, both views are deeply intertwined with each other; thus, it is important to incorporate them in a balanced manner. An organization may carefully analyze the positioning that should be taken when it enters a new market. It may also become sensitive to what is happening in every corner of the organization in order to find desirable practices once a plan is put into practice. Even so, the dichotomy creates a framework that understands the nature of strategy and prevents discrepancies in the following discussions of the role of accounting in strategic management.

2.3. The role of accounting in strategic management

2.3.1. *Transmitting a prescriptive strategy within an organization*

Traditionally, the role of accounting has been debated based on a prescriptive view of strategy. Anthony [ANT 65] positions accounting as a means of transmitting strategies formulated by senior management to shop-floor employees for the purpose of consistent implementation. The role of accounting is to compare periodically actual results against preset targets and let the users know the need to fill the gap. This kind of view on the role of accounting is called the "cybernetic model" [ASH 56] because the expectation is mechanical, like an air conditioner that maintains a room's temperature.

Simons [SIM 90, SIM 95, SIM 00] proposes a framework called the levers of control (LOC) to accomplish organizational objectives that address the uncertainties an organization faces. He insists on a dual usage of accounting as a "diagnostic system" and an "interactive system". The diagnostic use of accounting is almost equivalent to the cybernetic view and aims to monitor the outcomes that need to be accomplished as originally planned. At the same time, the diagnostic approach can be used interactively. Budgeting processes, for instance, give managers a chance to communicate with their subordinates. A gap between a preplanned target and its actual result is not a reason to accuse employees but a clue to understanding changes in the business environment.

Together with these two styles of control, Simons [SIM 90, SIM 95, SIM 00] proposes "belief systems" and "boundary systems". A belief system is an organization's core value that indicates the direction an employee should take and clarifies the actions and behaviors he or she should adopt. A boundary system is a code for showing actions that should not be taken. By controlling these four levers in accordance with a situation, managers are expected to accomplish an organization's goals and address any uncertainties surrounding the organization.

Simons' works [SIM 90, SIM 95, SIM 00] have established a trend in accounting research called the "control package" [MAL 08]. The control package is a way of regarding management control systems as a collection of control systems, such as rules, practices, values and cultural controls. Managers are expected to manipulate these control systems to ensure that employees work to achieve organizational goals. Abernethy and Chua [ABE 96] demonstrate how the mix of control package is shaped and designed by strategic choices within an organization based on a longitudinal field study of a public teaching hospital in Australia. Marginson [MAR 02] concludes from empirical data at a British telecommunications company that the use of accounting in combination with belief systems promotes organizational learning. Further, Henri [HEN 06] claims that a balanced use of interactive and diagnostic performance measurement systems creates dynamic tensions within a company, a situation that promotes organizational capabilities by fostering organizational dialogue, stimulating creativity and focusing organizational attention.

The LOC framework and the following control package view have advanced strategic management accounting research. The foregoing discussions advocate a more active and flexible role for accounting than that represented by the cybernetic model. However, a major concern about the LOC framework, or the control package view, is the way in which senior management can obtain information at the shop floor level that indicates changes in the business environment. By utilizing the four levers of control, senior managers are expected to manipulate employees in terms of the strategy they have established in advance. In other words, senior managers see strategy as something they should formulate in advance, even though any uncertainties that inevitably require them to modify the strategy are to some extent assumed during the strategy's implementation. In addition, although the aforementioned studies emphasize a combinative use of accounting with other management control systems, they do not focus on the role of accounting itself in strategic management, apart from advocating accounting's interactive use.

Kaplan and Norton [KAP 96, KAP 01] propose a performance measurement framework called the balanced scorecard (BSC) in order to visualize a strategy by putting it into relationships of numerical indicators. They suggest that an organization and its strategy should be regarded from four perspectives: "learning and growth", "internal process", "customer" and "financial". They insist that an emphasis on accounting and financial indicators, such as profit or return on equity, leads to unbalanced management because these indicators merely represent past achievements brought about by a strategy's implementation. It is important to capture current and future situations by quantifying them from other perspectives as leading indicators. By drawing a "strategy map" that shows the relationship between indicators for each perspective, a strategy is visualized so that employees can understand the whole story of how the strategy creates values and financial outcomes.

Imagine a scenario whereby a high-class, stand-up beefsteak restaurant is opened within a narrow area in a city center. Since it is a challenge of developing a new position within the restaurant business, it may be regarded as a focus strategy referring to Simons' works [SIM 90, SIM 95, SIM 00]. Although a specific profit target is provided, the target is simply the figure that will be achieved as a result of the strategy being realized. Thus, what is needed is a strategy map that shows causal relationships between the four perspectives in a quantitative manner.

In order to provide high-class steak, the chefs need to have decent cooking skills and a knowledge of beef. Consequently, the restaurant may spend time providing the chefs with appropriate qualifications or recommending them to acquire such qualifications. In order to operate efficiently in a limited space, the chefs may also need to be given multiskill training so that they can clean the kitchen and handle money. These issues regarding the "learning and growth" of employees can be assessed by nonfinancial measurements such as training time and the number of qualified people.

The next point to consider may be the restaurant's operation. It may be inevitable that the material cost ratio deteriorates because of the use of high-class beef. However, the strength of a stand-up restaurant is in the high turnover rate of customers. By providing a meal on one plate, it may be possible to limit the number and types of dish in order to reduce the time taken for serving and washing. The customers may help themselves at a salad buffet while waiting for their steaks to be grilled. The number of staff at the restaurant will be limited because the chefs handle various tasks. There may be only a few items apart from beefsteak on a menu to simplify operations and reduce inventories. Most issues related to these "internal processes" can also be quantified through nonfinancial indicators.

Providing delicious and voluminous steak at a reasonable price for a quick lunch may result in a high customer ratio of single people in their twenties or thirties working in nearby offices. The cleanliness of the restaurant (which is probably an "internal process" factor) may also result in raising the female customer ratio. Since the restaurant is for daily use rather than, say, a special meal for anniversaries, the frequency of visits by regular customers will be higher than standard steak restaurants. Moreover, average sales per customer will be lower than standard steak restaurants but probably higher than standard fast-food or stand-up noodle restaurants. These factors related to "the customer" can also be quantified and used in a strategy map.

Such financial and nonfinancial indicators are interrelated. Improving the indicators that are related to the learning and growth of employees' results in better internal-process indicators. Better processes increase customer satisfaction and other related indicators and ultimately cause improvements in financial indicators. In this way, BSC ensures that strategic understanding reaches every corner of an organization by incorporating and embodying an abstract strategy into a strategy map based on causal relationships between financial and non-financial indicators.

BSC was initially regarded as a tool to transmit an organization's strategy and evaluate individual performances. In other words, it was based on the prescriptive view of strategy. However, later studies have focused on a further aspect introduced by BSC utilization. Hansen and Mouritsen [HAN 05], using four cases of Danish companies, illustrate how BSC helps to clarify problems inherent in each organization in order to encourage interest in such problems among employees. In the case of a pharmaceutical company, BSC helped employees to understand the necessity of cross-functional integration by showing how much other divisions were involved in the performance of their operations. In the case of a textile company, BSC introduced comparability between sales divisions. Because the company had grown rapidly, each division had developed its own style of operational processes. The unification of operational processes introduced by BSC enabled each division to benchmark against each other. This benchmarking resulted in operational improvements across the company. Hall [HAL 11] regards BSC as a comprehensive performance measurement system in which the relevance between the activities of individuals and the final organizational outcome is presented in a concrete manner. Understanding daily activities in a strategic context helps organizations achieve efficient operations. Such understanding even enables employees on the shop floor to find contradictions and problems inherent in management control systems and corporate-wide strategy.

The foregoing studies indicate that the selection of accounting and other management systems is not by itself directly linked to a specific type of strategic view. It is how the selected systems are used that determines the role of accounting in strategic management.

2.3.2. *Fostering organizational learning as a postscriptive strategy*

Despite the recent theoretical development of strategic management accounting, the criticisms of management accounting, based on a traditional view such as a cybernetic model, persist. They argue that management accounting systems exert their effects only in highly controlled environments. In a highly varied environment, such systems have negative effects such as short termism [HOP 72], the creation of slack [BRO 85] and suboptimization [MER 98].

At one extreme, Hope and Fraser [HOP 03] suggest abandoning a budgeting system in a highly competitive and changing market and to aim instead for "beyond budgeting" management. Alternatively, by reexamining the strengths and weaknesses of a budgeting system, some studies show the possibilities for accounting to engage in active roles, such as creating opportunities to encourage employees' creative ingenuity in order to respond to changes in the business environment.

In contrast to the concept of "beyond budgeting", Libby and Lindsay [LIB 10] show that 79% of North American companies have implemented budget management in order to motivate managers and evaluate performance. In addition, approximately half of the companies (44.0% in Canada and 51.2% in the United States) responded that no budget revision was implemented during the year. In such a context, it is worth considering why many companies still use a budgeting system with fixed targets despite a great deal of research highlighting its negative effects.

In a budgeting process, it is necessary to forecast the future of each division or organization in terms of accounting numbers. However, a varying degree of unexpected situations will occur during the period of the forecast. Revising the budget each time, the situation changes can improve its accuracy but result in organizational exhaustion [OTL 99]. Thus, it is important not to pursue the perfection of budgeting and accounting controls.

The notions of "strategy as practice" [WHI 96] and "practice turn" [SCH 01] have advanced strategic management accounting research based on the postscriptive view. These notions contend that strategy practitioners are not only senior managers but also many others who engage in strategic work. Strategy practices entail routines of behavior, norms and procedures for thinking that are implicitly shared in an organization. Strategy lies in the microprocesses of an organization rather than its content (e.g. which positioning should be taken and what kind of products the organization sells).

Drawing upon their notion of practices, Ahrens and Chapman [AHR 04] illustrate the case of a restaurant chain company to show how accounting control can support employees when they address inevitable contingencies at work. Although the company's manuals determine the amounts of materials and the standard costs for each menu in detail, the local managers of the restaurants do not follow the instructions to the letter. For instance, they adjust the quantities of desserts in order to prioritize customer satisfaction. It is possible that the accounting information, such as the standard costs for each item and the prices on the menus, gives the local managers clues so that they know how to enable the adjustment of desserts by arranging the mix of other dishes. At the case site, accounting controls do not limit actions but enable employees to take flexible and creative action. Accounting information, such as budgets and standard costs, deteriorates from the moment it is determined. Such information needs to be updated and "repaired" [ADL 96] as time passes as a result of the daily activities at the shop-floor level. The case company understands this limitation of accounting and uses it in a flexible manner.

Frow *et al.* [FRO 10] discuss the importance of continuous budgeting based on the case of a multinational company managing its document technology and service business. The company determines a fixed budget target at the beginning of each financial term. It is not just a target to be achieved but is also a measure to know what is happening inside and out of the organization by continuously comparing with the actual results. The continuous "in-process management" creates an opportunity to explore various options in order to achieve the fixed target. In searching for and selecting between various choices, the core values of the organization play an important role. Each divisional manager is given broad discretion regarding the means to achieve the budget target. They each try to solve the problems by discussing and interacting with not only divisional employees but also employees in other divisions. The managers decide what options should be taken by referring to the values and norms shared within the organization. In order to evaluate their performance, the emphasis is placed on qualitative analysis about why such a situation occurred and not just analysis from a quantitative perspective. By so doing, the company utilizes budget control not just for short-term accountability but also for a strategic fit to the business environment.

2.4. Making an organization literate about strategy

The recent trend to search for accounting's role regarding strategic management and organizational learning can be explained as an expectation that accounting should make employees literate about the strategy of their organizations. Rather than promoting strategy that is preplanned by senior management, it is necessary to encourage the abilities and possibilities that lie within an organization in order to manage a continuously changing environment in the direction along which the organization is aiming to proceed.

According to Bay *et al.* [BAY 14], the notion of "literacy" is divided into the following two concepts. The first is a concept to capture literacy as a technical ability typified by reading and writing. In accounting, this ability refers to the proficiency of calculations and knowing about related terms and concepts. The second concept is to see literacy as the ability to create meaning in accordance with individual local situations and contexts. Further, Nonaka and Toyama [NON 07] differentiate between technical knowledge and practical wisdom. Practical wisdom, or "phronesis" in Greek, is the ability to determine and undertake the best action in a specific situation in order to serve the common good. Since literacy as wisdom is not uniform or solely absolute, multiple literacies can coexist, depending on the situation. The decision of a restaurant manager regarding the extent to which he or she should rigidly follow the manuals and standard recipes, shown in Ahrens and Chapman [AHR 04], is an example of this kind of literacy. In this instance, "customer satisfaction", as a shared value, requires the ingenuity of a local manager with the overall support of accounting information. The manager creates his own meaning on each occasion of customer satisfaction and meets the accounting requirement.

The way to raise each aspect of literacy also differs. Literacy as knowledge can be transferred and copied from a literate person to an illiterate person. In order to do so, literacy as knowledge is expected to be transmitted accurately in explicit forms. With regard to literacy as wisdom, it is necessary to support learners' processes of creating their own meanings in accordance with circumstances. Nonaka *et al.* [NON 14] propose a notion of "dynamic fractal organizations" in order to foster practical wisdom throughout an organization. It is not only important for senior managers to pursue the "common good" through their span of control; it is also important to provide a chance for employees at every layer, including the shop floor level, to do the same.

In order to expand this line of argument, one can refer to cases of Japanese management and accounting practices. Okano and Suzuki [OKA 07] explain that a feature of Japanese management accounting is that it emphasizes communication between senior management and shop-floor workers rather than central control. The authors state that senior managers in Japanese companies tend to encourage the wider participation of shop floor workers in continuous improvement processes such as Kaizen costing. Further, referring to Japanese management and accounting practices such as those of Toyota, Johnson [JOH 92] criticizes top-down controls via accounting information and emphasizes the importance of bottom-up empowerment.

Cooper [COO 95] examines the widespread use of accounting controls in Japanese companies, calling them micro-profit center (MPC) systems. Small groups in production lines are not only evaluated by profit but also expected to budget and plan their activities yearly, monthly and, in some cases, daily. The accounting numbers used in each group are simple so that shop floor workers without accounting expertise can manage them on their own. Cooper [COO 95] argues that MPC gives group members opportunities to consider their roles in companies by showing their contributions in a quantitative manner. This approach is expected to raise the group members' cost and profit consciousness. However, the notion of MPC proposed by Cooper [COO 95] relates relatively to the technical aspect of calculative practice.

In Chapter 3, an in-depth case study of a Japanese manufacturing company is provided in order to understand further the possible role of accounting to support organizational learning. At the case site, accounting has an important role to play in raising members' literacy about the values and perspectives shared in an organization. In this regard, the widespread use of accounting provides an opportunity to consider the common good and the value-creating behaviors situated in the context of an organization's every corner. Contrary to traditional debates about the controllability principle in responsibility accounting, the widespread use of accounting increases divisional interdependencies. However, the case company focuses on facilitating communication, discussions and organizational learning via simple accounting numbers.

Part 2

The Linkages between Accounting and Strategy Practices in Various Environments

3

Management Accounting Practices as Organizational Learning: Continuous Value Creation in a Japanese Company[1]

3.1. Introduction

This chapter explores how management accounting systems and their practices contribute to continuous value creation by encouraging organizational learning based on an in-depth case study of a Japanese manufacturing company. Organizational learning has been an important issue in recent strategic studies as previously discussed in Chapter 2. Raising organization members' capability contributes to increasing long-term competitive advantage and profitability.

The company used in this case study, Kyocera Corporation, was established in 1959 as a small town factory. The first year sales were 26 million yen (or US$ 72 thousand at that time), which has rapidly grown to 1.479 billion yen, or 13 billion USD at the current rate, for the fiscal year ending in March 2016 (consolidated). During that time, the company has grown continuously for 57 years, without running a deficit. This study examines in particular its unique management and accounting system, called the Amoeba Management System (AMS), as a key property for continuously increasing its competitive advantages.

1 This work was supported by JSPS KAKENHI, grant number 16K17216.

AMS has become prominent following Cooper's [COO 94, COO 95] study of the factors that brings long-term value and competitiveness to Japanese firms. Selecting five Japanese firms that implement lean management and accounting practices, he finds that these firms are distinguished by their creation of numerous micro-profit centers. He dubs these practices the "MPC system", and cites AMS as one of the practices.

This chapter also references the strategic concepts and discussions of management accounting noted in Chapter 2 while analyzing the AMS case. Specifically, this chapter uses the concept of "inscription" from Robson's [ROB 96] study to examine the roles of a unique profit measure and its reporting format, known as the Hourly Efficiency Report (HER) at the case company. In AMS, organizational units (called "amoebas") classified by functions such as sales and manufacturing are each positioned as individual profit centers. Furthermore, management practices based on HER are observed at every level of the organization.

The leader of each amoeba is expected to set a profit target including related numbers (called a Plan), primarily on annual and monthly bases that incorporate the future he or she envisions in numerical terms. In other words, in this process, the leaders are not simply providing objective predictions of future performance; rather, this process requires that they "inscribe what you want to accomplish this month in HER" (a comment from an interview).

These Plans and their actual results require approval and analysis through discussions with supervisors (upper managers). Amoeba leaders are expected to understand and account for every figure in the HER, as well as the background circumstances for each figure. They are also expected to be able to explain this information to others by demonstrating feasibility in a concrete manner.

These practices based on HER appear at every level of the organization, regardless of the scales and functions of the amoebas. Therefore, the experiences that leaders gain at the lower levels of the organization can also be applied when they reach higher levels of management. In practical terms, only a few issues can be managed at the lowest levels. However, it is hoped that by gradually managing a wide range of amoebas at a higher level, leaders will reach a greater level of management and accounting literacy.

In the following, this chapter will first review the related literature focusing on management and accounting practices. After outlining the methodological issues and research method, we will provide a case description followed by its theoretical discussions and conclusions.

3.2. Literature review

A “bottom-up” approach to management and “total participation” have been cited as two distinguishing characteristics of management and accounting practices in Japan. For example, Okano and Suzuki [OKA 07], using Toyota as an example, cite the practices of total quality control/management (TQC/M) and continuous kaizen (improvement) activities as long-term value creating behaviors in classical Japanese management and accounting practices. Kagono *et al.* [KAG 83] identify “group dynamics” or “decision-making through direct interaction between individuals focused on a group structure, as well as the process-centric models that employ it” as a distinguishing characteristic of Japanese management. They also state:

> “Here (in Japanese management), total participation by average individuals results in reduction of diversity, but the problem (when it arises) is shared (in the entire organization) rather than divided. It will be gradually solved, sometimes by covering each other, or competing with others on other occasions” [KAG 83, pp. 115–116].

Cooper [COO 94, COO 95] takes case studies of five Japanese companies conducting lean management practices and identifies a shared characteristic for “creating micro-profit centers”, which he dubs the “MPC system”. This practice aims to increase the willingness to participate and the awareness of the firm’s profits among shop-floor workers by conducting profit calculations in organizational units consisting of dozens of workers. In contrast with strategic business unit management or in-house company systems, Cooper [COO 94, COO 95] identifies the small size of these profit centers as the distinguishing characteristic of the MPC system and cites AMS as its representative example.

While Cooper [COO 94, COO 95] focuses on an external characteristic, namely the creation of numerous micro-profit centers within an organization, Japanese researchers attempt to clarify the mechanisms by which the MPC system functions in a desirable manner. Some of these attempts reference the concept of “empowerment” [BLA 96] focusing on the widespread use of accounting information by shop-floor workers. In addition to “delegation”, which refers to the conferral of responsibility, the term "empowerment" is also used to convey a sense of "encouragement and cheering up".

For example, Tani [TAN 99] finds that the MPC is one total participation system that provides each leader with more opportunities to experience management to instill a sense of ambition in each work site. Tani [TAN 99] therefore concludes that profit measures that are "easy to understand, with a simple structure" serve as a "common language" by which all employees, regardless of their functions, can understand and use to communicate important issues at their daily activities. Matsugi [MAT 05] points out that the monthly meetings held at each MPC function as a cultural control beyond being an opportunity to simply communicate each division's situation. Watanabe [WAT 04] examines the effect of the MPC system from a psychological point of view. Based on a quantitative analysis, he concludes that the introduction of MPC increases leaders' intrinsic motivation by expressing their daily endeavors to improve daily activities in a positive numerical notation, which is profit, rather than the negative form, which is cost.

Researchers discuss the importance of simplicity or understandability of accounting measures in cases besides those of MPCs. For example, Horii [HOR 09] examines investment decision-making processes in a steel manufacturing company in Japan. Although it is often stated that methods such as net present value and internal rate of return are theoretically decent in accounting literature, Japanese companies generally tend to use the payback period method [RYA 02, ALK 06, SHI 07]. Horii [HOR 09] claims that an organizational context that stresses total participation decision making will emphasize methods that allow more organization members, including shop-floor workers without any accounting knowledge, to understand their results. Thus, the positive relationship between the simplicity and understandability of the accounting measures and the activeness of communication within the organization has been the subject of numerous additional studies, both inside and outside Japan [WEI 69, BRE 88, REM 93, KAZ 03].

Another important issue to understand value creation in the context of Japanese management is the role of management philosophy. For example, Hiromoto [HIR 09] explains that management philosophy is the key element to understand management accounting practices in Japan. He claims that employees' thorough understanding of the organization's management philosophy leads to organizational learning and creativity. Sawabe and

Tobita [SAW 08] quantitatively examine the relationship between management philosophy and corporate performance. Based on questionnaire responses from 167 companies in Japan, they find that the penetration of the management philosophy within the company has a positive impact on employee motivation and satisfaction, both of which have a positive relationship with return on equity.

Management studies have a long history of research into the fundamental visions and philosophy of corporations, one famous example being Ouchi's [OUC 79] examination of "clan control". In many cases, clan control is informal; however, it is sometimes also formal, taking such forms as credos and management philosophy [FLA 83, LAN 97].

As Alonso [ALO 87, p. 14] states, "there is no counting without concepts", in the sense that fundamental norms and concepts are inseparable in order to quantify an object. However, in the context of management accounting research, highly abstract management philosophy and the use of management accounting as a specific forecasting technology have so far often been discussed separately (refer to the discussions of levers of control [SIM 90, SIM 95, SIM 00] and related studies in Chapter 2). Cooper [COO 95] also refers to the role of management philosophy, but restricts himself to abstract explanations, such as:

> "The system prevented organizational bureaucracy by creating an environment in which all levels of management could interact freely for the common purpose of improving the firm's operations and pursuing its strategic objectives. The system works because of the corporate philosophy instilled by Kyocera's founder, Dr. Inamori" [COO 95, p. 306].

Research on balanced scorecard (BSC) aims to incorporate abstract properties such as corporate visions and strategy in concrete objectives in local organizations. By clarifying the concrete actions to take at each organization level as well as showing the causal relationships between these actions, BSC visualizes and transmits the abstract organizational goal to the entire organization [KAP 96, KAP 01]. Hall [HAL 11] treats BSC as a comprehensive performance measurement system and a means by which an organization's members can assign their own meanings to everyday situations.

Extending these arguments, the rest of this chapter explores the roles of management accounting for organizational learning as a means of continuous value creation. BSC research shows the possibility of visualizing abstract properties as numbers. However, its focus is on the role of non-financial numbers, which embraces local contexts at shop floors. Since the abstract objectives are broken down into non-financial numbers that are familiar to shop-floor workers, those numbers are easy for them to understand and let them know how they are engaging in the accomplishment of the corporate visions or strategy through their daily activities. On the other hand, relying on non-financial numbers may prevent them from understanding other local contexts since other non-financial numbers are probably used in that different context (especially in different functional divisions). In other words, each non-financial number simply describes each local context in different forms, rather than serving as a common language within the entire organization. This situation may create literacy that applies only to each local context.

Ahrens and Chapman [AHR 04] depict how restaurant chain managers realize an organizational perspective that emphasizes customer satisfactions in daily management accounting practices (refer to Chapter 2). They treat management accounting systems not just as a means of transmitting strategy but also for extracting the organization's strength and capabilities. Their study provides an important clue that clarifies the role of management accounting in fostering organizational learning in conjunction with subjective judgment based on corporate-wide visions and norms at the local level. However, the corporate vision in their case is relatively simple, that is, pursuing customer satisfaction. The case in this chapter will show more intertwined relationships between management accounting and philosophy, and further examine the role of management accounting systems and practices as an education system for continuous value creation.

3.3. Methodological issues

This chapter draws on the notion of "inscription" in [ROB 96] to analyze the case of Kyocera. Robson argues that the development of accounting should be considered in terms of a continuing refinement of mobile, stable and combinable inscriptions that expedite long-distance control. Inscription is defined as a material translation of unfamiliar events, places and people to be acted on.

Morgan [MOR 88] referred to numbers as the dominant metaphor of accounting. Quantity, which is expressed in numbers, is inseparably related to quality; as Alonso argues, "there is no counting without concepts"

[ALO 87]. Once the object of quantification is defined, the numbers then come to stand for the conceptual entities. Robson [ROB 96] claims:

> "[…] the content of a number statement, that is, a sentence which assigns a definite number to a set of objects, is an assertion about a concept rather than an object […] if, for example, I state 'I have nine cars', this can firstly be restated as 'the number of my cars is nine […]' in this form the cars now fall under the concepts 'car' and 'has nine instances', both of which are concepts in identity with themselves" [ROB 96, p. 688].

In this example, what is actively suppressed by the process of quantification is that "my cars" may differ in design, capacity, color, top speed and so on, because once particular concepts or properties of objects have been defined and counted, differing or distinguishing attributes are no longer visible. Ezzamel *et al.* [EZZ 04], for example, examine the role of accounting calculations as inscriptions that transformed and promoted the new commercial agenda in a high-tech division of a major British manufacturer. They focused on the increased and changing use of performance measures at this site, interpreting them as actions that signified and facilitated an increasingly commercial orientation toward activities.

Latour [LAT 87, LAT 88] states there are three major and interrelated components for inscriptions as the means of acting on unfamiliar events, places and people at a distance. Inscriptions need to:

1) render them mobile so that they can be brought back;

2) keep them stable so that they can be moved back and forth without additional distortion, corruption or decay;

3) be combinable so that regardless of the stuff they are made of, they can be accumulated, aggregated or shuffled like a pack of cards.

Robson [ROB 96] argues that these three elements are the qualities that numerical inscriptions also need to possess. In an accounting context, the mobility of accounting reports is strongly attached to the use of writing. Company accounts inscribe productive processes, labor, transactions, machines, buildings, cash flows and loans dispersed spatially and temporally but given a common form by the text. By the quality of stability, Robson [ROB 96] means that inscriptions must be recognizable to their users, implying the stability of the relationship between the inscription and the context to which it refers. Combinability is defined as what allows the actor to accumulate, aggregate, tabulate and recombine inscriptions to establish new relationships and calculate norms to compare the settings to be influenced in accordance with his or her specific objectives, aims or ideals. Robson [ROB 96] contends that accounting statements apply the characteristics of combinable inscriptions by giving us a chance to appreciate the process of translation through which this combinability is achieved. He further posits that accounts numbering the details of transactions, productive equipment, labor and so on do not discriminate between these entities, but assert that they have identical qualities.

3.4. Research design

The analysis in this chapter is based on an in-depth case study at Kyocera Corporation Co., Ltd. (Kyocera), a manufacturing company in Japan, and its group companies. The main products of the company are fine-ceramics and electronic components based on fine-ceramic technology. Within the Kyocera group, there are 235 companies, primarily based in Japan, the United States, Europe and Asia. As of the end of the fiscal year, Kyocera boasted 69,229 employees.

The research presented here consists of two parts. The first part is based on semistructured interviews that were conducted from May 2004 to February 2009. The series of interview are divided into three periods.

The first period (from May to September 2004) consists of interviews (four sessions, 9 h in total) of the head of management consulting at Kyocera Communication Systems Co., Ltd. (KCCS), a company of the Kyocera group, which operates an amoeba management consultancy both inside and outside the firm. The primary goal of this portion is to gain a complete picture of the amoeba management conducted at Kyocera and an understanding of the firm's fundamental structure.

The second period (from September 2004 to February 2008) consists of interviews (10 sessions, 20 h in total) conducted at Kyocera factories. The interviews focus on leaders at the shop-floor level, as well as the section-level supervisors and department-level amoebas above these leaders. Here, the primary goal is to collect qualitative data on practical knowledge on implementing amoeba management from a shop-floor standpoint, as well as to substantiate the interview data acquired in the first phase and identify points of difference.

The third phase (conducted in January 2006, and from October 2008 to February 2009) focuses on interviews (10 sessions, 15 h in total) concerning the introduction of amoeba management at external firms. The interviews were conducted with the KCCS presidents and consultants. The primary industries targeted were hospitals, telecommunications equipment manufacturers and business printing equipment manufacturers.

The entire interview research conducted over the above three phases was on a semistructured basis, and while a general question list was provided to interviewees beforehand, additional questions were posed as they suited the flow of the conversation.

The other part of the research is an archival analysis that has been conducted from May 2004 to the present. Both internal and published documents, including HER of amoebas, internal company magazines and books and Websites regarding the case company and AMS, have been collected and observed. It has been mainly conducted to confirm the validity of and supplement the interview data.

3.5. Case description: ubiquitous management accounting practices at Kyocera

3.5.1. *Organizational structure and accounting responsibility*

The organizational structure at Kyocera generally follows a pyramid structure. Business headquarters and units are created primarily by product, under which department for each function (e.g. sales and manufacturing) is located. Additional subordinate sections and teams are further organized as necessary. The sections and teams at the bottom of the organization consist of a handful of people on the smaller side and up to 50 employees on the larger side.

There was an emphasis on establishing sales, manufacturing and administration divisions separately. Companies with strong manufacturing divisions may have sales functions within manufacturing divisions, but from the perspective of clarifying the roles and responsibilities of each organization, AMS created separate organizations for these functions. The characteristic is that both the sales and manufacturing divisions are treated as profit centers (the administration division is treated as a cost center). Many of Kyocera's products are built-to-order. Product sales are recorded as part of the manufacturing division, not the sales division. The sales divisions serve as a liaison between the customers and the manufacturing division, that is, to communicate customer desires to the manufacturing division and to communicate the value of products to the customers and encourage purchases at reasonable prices. The manufacturing divisions pay a set percentage (about 10%) of product sales to the sales divisions as a commission fee, which counts as revenue for the sales divisions.

Further, the lower level sections and teams (amoebas) are also treated as profit centers, and each leader is expected to serve as management and to keep his or her individual unit (amoeba) profitable (however, if the sales and manufacturing divisions have their own administration department within them, they are treated as cost centers). The individual units at each level of the organization are called "amoebas". The units in the sales divisions are called "sales amoebas", and the units in the production division are called "production amoebas". Amoebas positioned as profit centers and cost centers are called "profitable amoebas" and "non-profitable amoebas", respectively. There are also "department amoebas", "section amoebas", and "team amoebas", labeled depending on their level. Kyocera alone employs approximately 14,000 employees, but there are about 3,000 amoebas at all levels [INA 13].

The sales division forms sections and teams for each product and client; therefore, these units' daily activities are relatively independent. In the manufacturing division, however, multiple sections and teams are involved in the manufacturing of a single product. For example, there may be several manufacturing processes for a single ceramic product such as mixing raw materials, molding, firing and grinding, each of which different amoebas are responsible for. When an order is received from a customer, the related amoebas gather to estimate the amount of materials and working hours required to fulfill the order and discuss the amount of profit each amoeba will generate within the suggested amount of money or what ingenuity is required to generate profit within that range. Thereafter, based on that discussion, the sales received from the client are distributed among each amoeba. In this structure, both the sales and manufacturing amoebas, including the lower levels of the organization, are continually expected to act with an awareness of market prices.

Furthermore, each amoeba must account for their costs, including raw materials, depreciation, water supply and other costs allocated to the administration division; however, they do not include the labor costs for their members. Therefore, it is more accurate to say that each amoeba is counting their added-value. Alternatively, the amoeba will tally its members' working hours. The working hours in the administration divisions are also tallied and allocated to profitable amoebas. The total working hours of each profitable amoebas are thus calculated. The profit (added-value) of each amoeba is then divided by its total working hours to calculate hourly efficiency. Amoebas are allowed to lend and borrow personnel between them and working hours are traded between amoebas. Therefore, by lending members to amoebas that are short on workforce, an amoeba can reduce its total working hours and improve its hourly efficiency.

Combining numbers with different dimensions, profit and time to create hourly efficiency makes it possible to compare the performance of amoebas at different scales, functions and levels within the organization. If they are evaluated solely by the amount of profit, it is difficult to compare amoebas to different scales. If using non-financial indicators, such as the BSC, it is difficult to compare units of differing functions. It is also fractal in the sense that compiling the performance of each amoeba falls to an upper amoeba, ultimately making it possible to measure company-wide hourly efficiency. Personnel expenses are also tallied at the department level and above; if these expenses are deducted, then it is a simple task to calculate profits for external reporting (financial accounting) purposes using company-wide hourly efficiency.

3.5.2. *Penetrating the Kyocera philosophy*

AMS focuses on the unique management philosophy, called the Kyocera philosophy. Along with expressing this philosophy in writing, an effort is made at Kyocera to ingrain the philosophy throughout the organization. The Kyocera philosophy also encompasses the entire management philosophy of Kazuo Inamori, the founder of Kyocera, which encompasses a wide range of material, including the dozen books written by Inamori. The philosophy is often clarified, however, as a "Management Rationale" and the "Twelve Management Principles" listed below, which are often considered representative of the philosophy.

3.5.2.1. *Management rationale*

The following is an extract from the Kyocera Philosophy Pocketbook [KYO 94]:

> "To provide opportunities for the material and intellectual growth of all our employees, and through our joint efforts, it contributes to the advancement of society and humankind.
>
> Twelve management principles as follows:
>
> 1) clearly state the purpose and mission of your business;
>
> 2) set specific goals;
>
> 3) keep a passionate desire in your heart;
>
> 4) strive harder than anyone else;
>
> 5) maximize revenues and minimize expenses;
>
> 6) pricing is management;
>
> 7) success is determined by willpower;
>
> 8) possess a fighting spirit;
>
> 9) face every challenge with courage;
>
> 10) always be creative in your work;
>
> 11) be kind and sincere;
>
> 12) always be cheerful and positive. Hold great dreams and hopes in the pureness of your heart".

The Kyocera Philosophy Pocketbook, which details Kyocera's unique management philosophy, is distributed to all Kyocera employees, and at every daily meeting, employees have the opportunity to come together and recite the Kyocera Philosophy or express their own thoughts. Furthermore, Kyocera Philosophy Education (spanning two sessions lasting 2.5 h each) is conducted for all employees; in March 2016, 37,347 employees attended these programs. These study sessions are used as an avenue for employees to share with one another case studies and experiences where the Kyocera philosophy was used to make decisions in their daily work.

Much of the Kyocera philosophy, including the above tenets, is abstract. However, when one attempts to gain a more concrete grasp of the philosophy through related documentation and interviews, one finds that the philosophy is not always consistent; at times, it seems to contradict itself, espousing a variety of belief systems and approaches. For example, Inamori [INA 13] uses the term "extended-family principle" to explain thusly the fundamental relationship between members of the Kyocera:

> "If the company becomes a community held together by destiny, as one extended family, and if mutual understanding, encouragement, and help are freely given among managers and employees, just as in a family, then it should be possible to run the company with management and labor as one body united for the same purposes. Even in the face of severe market competition, operations should naturally proceed well, as united efforts are directed toward the development of the company. I named this concept the *extended-family principle* and made it part of the company's management foundation" [INA 13, p. 19].

In other words, this approach stresses, above all, a familial spirit of mutual aid. The worldview professed by the "extended-family principle" is also connected to the computation structure of hourly efficiency. As detailed above, the index by which each amoeba is evaluated is not total profit itself but rather hourly efficiency, which is added-value divided by total working hours, with personnel expenses excluded from costs. Each amoeba is expected to follow the "maximize revenues and minimize expenses" guideline from the above "twelve management principles". However, since personnel expenses are not treated as costs, they are not the target to be minimized. Instead, due to the rule that allows working hours to be traded between amoebas, workers are proactively sent to amoebas that are short on manpower; it is thus expected that amoebas will help each other. This sort of worker movement is described thusly by the leader of a shop-floor level amoeba in an interview:

> "Our people, for instance, you know, they are scheduled to go here and there at the beginning of the month, at our planning stage. For example, this month, at the shipping stage, if personnel numbers go up or down by one or two people, that is often. …That happens like every month".

On the other hand, as expressed by such words as "you make your own bread", [HAT 15] each amoeba is also expected to be self-reliant. The following interview comment was offered in connection with this approach:

> "Our idea is that every amoeba should be in the black. It's not acceptable for them to be in the red. They have to think about how they can be in the black. Is it an issue of pricing, or is it an issue of management? ...If you look at the big picture, it is rarely the case that it is tolerated to be in the red" (September 21, 2004, Head of Business Management Consulting Office)".

Spiritual and idealistic principles are also at work in the Kyocera philosophy, as evidenced by their management principles: "keep a passionate desire in your heart", "success is determined by willpower", "possess a fighting spirit", and "always be cheerful and positive. Hold great dreams and hopes in the pureness of your heart", among others as shown in the aforementioned twelve management principles. On the other hand, as professed by the directives "Move steadily forward one step at a time", "Take a look at yourself", "Pay attention to the details", "Recognize what you cannot do" and "Obtain and challenge yourself with support" in [KYO 17] and [INA 01], pragmatic belief systems focused on stability are also present, directing members to be aware of reality, establish clear goals rooted in the facts, and deliver solid results.

In this manner, even though the term "philosophy'" is used, many different belief systems and approaches are embraced at Kyocera, and it is clear that these engender contradictions and clashes of beliefs.

3.5.3. *Budgeting processes based on Plans*

At Kyocera, budget control is conducted primarily in two cycles, yearly and monthly, based on an hourly efficiency target called a Plan. Yearly budget control is directed in basic annual policies by the head of each business office. Under these policies, a set of goals (Plan) is established based on more detailed figures focused on the department level. A Plan for each unit is created using HERs. To create the Plan for each department, hourly efficiency targets are set for each lower level amoeba (section and team). The budgeting process then follows, involving every member of the company.

These annual Plans are also applied to the monthly Plans for each amoeba, which settles on a Plan by the first working day of the month; on the first of each month, a meeting is held to study the validity and appropriateness of these Plans at every level of the organization. The monthly Plans, then, are created in a bottom-up style. Though directives and other instructions are received from the supervisor, it is presupposed that the monthly Plans will ultimately be judged and decided by the amoeba leaders.

These planning processes are detailed in Miya [MIY 03], which also took Kyocera as a case study:

> "First, four sections belonging to each department (two machining sections, one firing section, and one grinding section) write up 'Previous Month's Production Plans', 'Previous Month's Gross Production', 'Previous Month's Hourly Efficiency', 'Current Month's Production Plans', 'Current Month's Expense Plans', 'Current Month's Added-value', 'Current Month's Total Working Hours Plan', and 'Current Month's Hourly Efficiency' on a whiteboard, based on Plans prepared in advance by each section.
>
> The performance data are transmitted at the end of the month from the administration division to each manufacturing amoeba. The production Plans are established based on backlog information from the sales of each product. After the totals for the four sections were calculated, the hourly efficiency for the entire department was 5,870 yen.
>
> However, the department supervisor stated that 'the hourly efficiency for the previous month was 5,950 yen; therefore in order to meet the yearly plan, we want this month's hourly efficiency as a department to be 6,000 yen'. The supervisor said, 'we want everyone to think of a way to somehow raise our hourly efficiency by 130 yen'.
>
> Examining the content of each data item revealed that overtime was on the rise. The department supervisor communicated that 'if you just cannot do this, let me know, but I want to adjust the section workloads so that overtime is kept to six hours per person'. Each section revised its working hour Plan on the whiteboard.
>
> Multiplying the working hours by 6,000 yen hourly efficiency, the amount of added-value to earn to meet the target was determined. When this was compared to the added-values each section submitted, it was realized that they had to compensate for a 1.2 million yen shortfall.

> This 1.2 million yen of added-values could be recovered by increasing production or cutting expenses. All participants checked the details of expenses and of the orders that had been accepted. Then, one of the machining sections promised to "increase yield rate and decrease expenses by 200,000 yen". Further, the firing section said, "we'll try to increase production by 1 million yen. However, the estimates look like we are going to reach 500,000 yen; we do not yet know where we will get the remaining 500,000 yen. We'll be reporting in again with our progress mid-month". Therefore, it was decided to go with hourly efficiency target of 6,000 yen for November" [MIY 03, pp. 106–107].

Each amoeba establishes an hourly efficiency Plan by the first working day of the month, going through the process such as that described above. The expenses for each amoeba are checked in detail, item by item, by superiors. If the expenses generated by administration divisions (non-profitable amoeba) escalate, the neighboring amoebas demand a reasonable explanation because for them it results in an increase in allocated expenses.

The results are tallied on a daily basis and communicated to each member at an assembly the next morning, with items deemed to be of particular importance given priority. A fiscal deadline of 3 pm on the last day of the month is set, and the monthly performance results must be tallied by the next working day. Performance evaluations compare the Plan with its actual result and perform a comparative analysis using hard data. The evaluations emphasize the reasons behind any discrepancies rather than simple numeric comparisons. Repeated discussions are held within the amoeba or between upper management and the leader of the relevant amoeba using the HER with preplanned estimates and actual results of the month.

Personnel evaluations do not employ a performance-based compensation system based on short-term results. For example, in evaluating a leader, the results achieved over the past several years by the amoeba the leader manages are taken into account, and a long-term, comprehensive evaluation is created. Inamori [INA 13] compares this to a performance-based compensation system, saying, "this method appeals directly to the worldly desires of employees", expounding this as follows:

> "Under Amoeba Management, there is little difference in individual incentives based on short-term yields. However, for individuals who work hard for the benefit of everyone else and

show good results in the long term, ability is duly assessed and reflected over the longer term in wage increases, employee bonuses, promotions, and so on" [INA 13, p. 38].

3.5.4. *Fractal organizational learning through the budgeting process*

The hourly efficiency data for each amoeba are assigned to the respective amoeba's superiors (the department or the entire business unit) and are ultimately summed up to report management results for the entire company. It is crucial that the accounting data be frequently and ubiquitously used at every corner of an organization in AMS. Even a small amoeba at a lower level uses the HER to set a monthly and yearly Plan and to tally and analyze its performance results on a daily basis.

The numbers used for each amoeba are detailed down to the 1 yen and cover a wide variety of items. However, items such as utility expenses, delivery charges, materials purchase and shipping costs are not individually difficult to understand. All members of each amoeba have a discussion to estimate how much they will be incurred on a yearly, monthly and daily basis. When a member becomes a leader, this system fosters an understanding of gradually more advanced content through activities connected with the management of increasingly larger organizational units, such as the determination of shared expense allocations with other amoebas.

The monthly meetings held at every level of the organization to formulate plans provide particularly crucial leader education using the HER. For example, as in this interview comment:

> "We prepare HERs that incorporate everything into their data: people, equipment, the works [...] When we are thinking about our Plans, we discuss with the company president at the monthly meetings about things like whether we are diverging from the yearly plan, and if we are, what to do about it. And at the meetings, we have things pointed out to us – like, well, here, look at this, this is strange [...] If we cannot (ultimately) persuade (the managers), our figures will not be approved" (July 27, 2004, Head of Management Consulting Office).

Inamori [INA 13] also expresses similar thoughts:

> "In cultivating leaders, appropriate guidance and assessment of each unit's management by senior management (including top management) are important parts of the process. I have been using meetings as places for conducting this kind of practical education. Executive meetings and other management conferences are venues for individual amoeba leaders to present the results of the previous month and plans for the next month based on the hourly efficiency report. Through the content of the presentation and the ensuring discussion, the cultivation of talent takes the form of frank appraisal and guidance on each leader's way of thinking and attitude toward the work" [INA 13, p. 121].

Furthermore, the budgeting process is also positioned as an avenue to implement the Kyocera philosophy. As expressed in the comments below, planning not only consists of objective performance prospects, but is also a process that takes "passionate desire" into account and incorporates an approach that emphasizes "what you want to accomplish" into the HER:

> "As the person (leader) responsible for your division, you made 3,500 yen (Hourly Efficiency) last month. Now, you need to inscribe what you want to accomplish this month in HER as your Plan. Let's say the options are 3,000 yen or 3,500 yen or more (as the Hourly Efficiency target of this month). Well, if you choose the higher, it is fine. But if not, and if your supervisor is expecting, let's say, 3,600 yen in the mind, the person (supervisor) may say 'think this over again'. However, (the supervisor) won't just say 'it has to be this number'. Supervisors say what they need to say, and vice versa. Then, they may agree to what the leader is saying, or may not. Through such a discussion, you may say 'I will do my best, so let me try for 3,550 yen this time' in the end" (as the amoeba leader) (July 27, 2004, Head of Consulting Business Office).

> "We are told through meetings that we need to have high goals and cannot just settle for what is already within our reach when making a Plan. It is important to challenge ourselves to reach high goals" (September 13, 2004, Administration Department Manager).

> "In our case, Planning is everything; therefore, our Plans are extremely important because we cannot do better than the Plan.

> We have always taken a stance of setting the highest target we can achieve and sticking to them" (September 21, 2004, Manufacturing Division, Department Manager).

On the other hand, each amoeba leader is expected to set realistic Plans that adhere to concrete pre-planned actions – just as the Kyocera philosophy expresses "move steadily forward one step at a time", "pay attention to the details", and "recognize what you cannot do", which emphasizes steadiness and reliability. By doing so, each leader is expected to clearly show the feasibility of the Plan to others. Regarding this, Inamori [INA 13] states, for example:

> "Having prepared the monthly plan, the amoeba leader must pass details of the program to amoeba members and ensure that targets are well known and understood. For amoeba members, making the targets well known and understood equates to making the goals their own. The goals need to be shared by all members to the extent that each person, if asked, is able to recite this month's estimates for orders, sales, production, and hourly efficiency, among other things. Building on that, concrete action plans for achieving the estimates are broken down into targets for individual members" [INA 13, pp. 106–107].

In this manner, each amoeba leader must not only incorporate "passionate desire" and an "ardent fighting spirit" into the HER in the form of numeric data, but must also "pay attention to the details" and "move steadily step by step" when doing so. However, making such a Plan is not easy. Lower level amoeba leaders in particular have a rather narrow scope of control; therefore, proactively seeking the support of other amoebas and gaining additional information and cooperation is essential. For example, take the following comments from an interview at Company A, which introduced AMS based on KCCS consultations:

> "Before introducing Amoeba Management, there was almost no communication between divisions. I mean, there was a sense of distance. However, when Amoeba Management was introduced, the divisions could not accomplish anything without talking to each other. Now, such communication is taken as a given. Since everyone exchanges their own opinions, sometimes there are disputes" (January 18, 2006, Production control manager at Company A).

> "It is particularly true for young people and part-timers. When making Plans, members would be proactive in going to other amoebas to discuss. There have even been incidents where by the time an important issue has become known, the part-timers have already enlisted the help of full-time employees to settle the matter" (January 18, 2006, President of Company A).

As stated above, the Kyocera philosophy encompasses numerous beliefs systems and ideas, some of which contradict and oppose each other. However, these are highly abstract, general contradictions, and the act of reading them alone poses no direct dilemma for members of the organization. However, in order to apply and implement the tenets of this philosophy on a daily basis and in a concrete manner, it is clear that members extend themselves and reach beyond the boundaries of their own control; they are creative and proactive in enlisting the help of others to overcome the difficulty. The ubiquitous use of the HER spread the process to every corner of the organization.

3.6. Discussions and conclusions: management accounting as an education system

Historically, management accounting originated as a method of managing factories and other facilities which are situated at a distance from the company headquarters [KAZ 89]. Conforming to the notion of "inscription" [ROB 96, LAT 87, LAT 88], conditions at remote facilities were originally converted into numerical but non-financial data, such as order numbers, working hours, production quantities, machine operation hours and yield rates, which were "stable" in the sense that they remained unchanged across time and locations. In other words, whereas such numerical data excludes contextual contents that can only be qualitatively expressed in the form of sentences, it "mobilizes" the conditions of the factory without decaying as time passes or traveling remote distances. Furthermore, by combining and converting these numerical data into a monetary basis, such as sales, costs, profits (as a differential amount of sales and costs), assets, liabilities and capital (as a differential amount of assets and liabilities), it became comparable between distant places and across times.

Extant literature (e.g. [OKA 07, KAG 83]) identifies the value of Japanese management as bottom-up approach and group dynamics. This chapter further explores the role of management accounting in that context for continuous value creation. Based on the notion of "inscription", the in-depth case study describes the diffusion of management accounting practices throughout the organization. The profit index (hourly efficiency) is applied to a manufacturing division, which is often treated as a cost center in other companies. Product sales are directly recorded in a manufacturing division; therefore, the HER inscribes not only the situation of its production activities in the form of costs, but also the market's evaluation for the activities in the form of sales amount. Consequently, each amoeba is expected to be continually aware of the market conditions, too.

Such management practices based on hourly efficiency are also observed among lower amoebas consisting of several employees. Product prices and the allocation of sales between amoebas are collectively determined by the amoebas involved in production. They also share some overhead expenses. Given these conditions, situations do occur where lower level amoebas have little control in their management decisions. Nevertheless, the important point is that hourly efficiency is used as a common index at every level in every functional division within the company.

During the formulation of monthly and yearly Plans, supervisors expect the leader of each amoeba to explain various items found in its HER. As implied by questions and comments, such as "what do you want to accomplish this month (as a leader)?", and "we need to have high goals and cannot just settle for what is already within our reach", leaders are asked if they are inscribing their own interpretations and translation of the Kyocera philosophy in their Plans of the HER. At the same time, these leaders are also asked if they inscribe its reality and feasibility in the HER supported by concrete action plans. At monthly meetings, amoeba leaders discuss all stable data items inscribed in their HERs with their supervisors, as well as the reciprocal relationship between the inscribed Kyocera philosophy and practical plans of action, thereby contributing to the construction of literacy as wisdom for their daily management.

Employees accumulate experience as end-unit amoeba leaders, eventually growing to lead higher level amoebas and gaining experience in management at a broader scope. AMS provides an instructive processes for lower level amoeba leaders to foster organizational learning. In this vein, the management accounting system and its practices serves as a fractal education system for continuous value creating behaviors in our case study of this company.

4

New Ways to Create Value Integrating Strategic and Accounting Issues in a New Context of Digital Transformation: French Perspectives

4.1. Introduction

The former chapters (particularly Chapters 1 and 3) have focused on value creation and linkages between strategic management and accounting issues. This chapter is dedicated to the evolution of strategic concepts in contemporary debate. How have they evolved over time, especially in the context of digital transformation? Do they take into consideration accounting issues and vision?

Among other consequences, the digital transformation reshapes value chains, business models and more broadly business practices (including accounting activities). In addition, advanced analytics tools ("business analytics" or "big data analytics") allow an access to infinite data and could lead to the adoption of business intelligence (BI) strategies for many kinds of public and private organizations. Advanced BI tools thus contribute to value creation for the organization and, in addition, an incremental transformation [LOE 15].

This chapter is organized as follows. In section 4.2, we discuss the evolution of various strategic notions (value chain, business model). In section 4.3, we show how digitization should impact accounting practices and activities based on the French situation and perspectives.

4.2. The evolution of value chain and business model

In Chapters 1 and 2, we have already presented the well-known value chain (at company level) developed by Porter in the 1980s. We will focus in this chapter on the value chain at the industry level as it also raises several questions (Tables 4.2 and 4.3) in relation to cost analysis. Regarding the notion of a business model (BM), it has close links with the value chain. As defined by [CHE 02], the function of a business model is to "articulate" the value proposition, select the appropriate technologies and features, identify targeted market segments, define the structure of the value chain and estimate the cost structure and profit potential.

4.2.1. *The extended notion of value chain*

The framework of the value chain can be extended to an entire industry. The industry value chain is composed of all the value-creating activities within the industry ending with the completed product delivered to the customer.

The scope of this value chain extends beyond organizational borders and makes linkages between various players. This type of value chain is a very relevant concept for analyzing the behavior of firms and markets. Indeed, a firm's value chain is part of a larger system that includes the value chains of upstream suppliers and downstream channels and customers. Companies achieve a competitive advantage by managing the value chain better than other companies in their industry. Porter calls this series of value chains the value system, shown conceptually below:

... > Supplier Value Chain> Firm value chain> Channel value chain> buyer value chain

Linkages exist not only in a firm's value chain, but also between value chains. Reaching a sustainable competitive advantage requires that the organization understands the entire value delivery system, not just the portion of the value chain in which it participates. Through analysis of the value system, a company can identify potentials for strategic alliances with various actors in the industry value system. While a firm presenting a high degree of vertical integration is poised to better coordinate upstream and downstream activities, a firm having a lesser degree of vertical integration can nonetheless forge agreements with suppliers and channel partners to achieve better coordination. In addition, it is possible to identify cost drivers and linkages within a value chain.

If the value chain at company level gives some ideas about the positioning of a firm within an industry, it does not allow an analysis of the industry structure in which the firm competes. In addition, in most sectors, it is very difficult for a single company to perform all activities from product design, production of components and final assembly to delivery to the final user by itself. Most often, companies represent elements of a value system. Hence, value chain analysis should cover the whole value system in which the organization operates. The value system includes the industry value chain.

A representation of a value chain at the industry level is quite different from a value chain of a firm:

– support activities are no longer represented;

– it includes the names of the main market players and gives information to a certain extent about linkages between the actors in the industry but it does not show the strength of the actors and the evolution of linkages between them;

– it shows how the activities that comprise a value chain can be contained within a single firm or divided among different firms;

– it is closely related to the value network, a concept used to present various actors in a combination of several value chains;

– it presents a comprehensive understanding of the dynamics of each link along the whole value chain at multiple points in time.

The value chain concept is sometimes confused with other notions: supply chain, cost chain, value system, etc. Table 4.1 presents and discusses these notions useful for analysis of the competitive strength of the organization.

According to several authors, the value chain framework has close links with a cost analysis. Shank and Govindarajan [SHA 92] consider that the value chain is not just an abstract conceptual tool. They advocate the value chain as most useful for cost analysis and they show to use the value chain model to undertake cost analysis (Table 4.2).

Concept	Definitions – main characteristics	Level of analysis
Value chain (company level)	Value chain analysis describes the interlinked activities that convert inputs to outputs within organization. It evaluates which value each particular activity adds to the organizations products or services.	Company (strategic business unit)
Value chain (industry level)	An industry value-chain is a physical representation of the various processes involved in producing goods and services delivered by the key actors of the market. It has some links with the supply chain (see below).	Industry
"Cost chain"	The notion of "cost chain" is not common as the cost analysis can be conducted through the value chain by assigning costs to the value chain activities. Determining total costs could be a complex challenge. In some cases, this cost chain can explain the interactions between the evolution of costs and market prices. Grant [GRA 05] discusses this approach to cost analysis by using the value chain in the case of an automobile manufacturer: "The approach to cost analysis (…) – identifying cost drivers and exploring their impact on the different activities of the business - is a useful diagnostic tool, but tells us little about how companies actually implement cost-cutting measures" (p. 267).	Company /industry
Supply chain	The supply chain comprises the steps it takes to get a good or service from the supplier to the customer. Supply chain activities transform raw materials and components into a finished product that is delivered to the end customer. Supply chain does not have the same meaning as logistics: logistics refers to the distribution process within the company, whereas the supply chain includes many firms such as suppliers, manufacturers and retailers.	Industry
Value system	Porter called the network organizations involved in the production and delivery of an offering to the end customer a value system. Value systems integrate supply chain activities (including first-, second-, and third-tier suppliers). Value system is very often confused with the notion of *filière* (see below).	Industry
"*Filière* analysis" (vertical production chain, global value chain, commodity system)	The *filière* approach has been developed in the 1970s by French researchers and represents a tool of the French industrial economy's school. Morvan [MOR 85] considers a chain (*filière*) as linked operations for the transformation of a good: activities can be grouped into upstream activities	Industry

	(supplying and manufacturing), and downstream activities (marketing and distribution) in a vertical representation. The chains are influenced by technology and have complementary interdependences [BAT 01]. According to Raikes *et al.* [RAI 00, p. 15], "its main objective has been to map out actual commodity flows and to identify agents and activities within a filière, which is viewed as a physical flow-chart of commodities and transformations". This approach emphasizes the measurement of input–output relations and prices at different stages of the production chain. "The intention is to find (also) those in the group of actors who not only determine their own action in the filière, but also thus powerfully influence the ability or even the need of other actors in the filière to act" [LEN 93, p. 2]. Value chain is very often confused with the notion of *filière* but these two notions are quite different. The *filière* is a vertical approach and it is closely linked with vertical integration strategy, transactions cots and internalization theory. This is not the case with the value chain.	

Source: Developed by the author, based on analysis of the articles cited.

Table 4.1. *Value chain and related concepts*

4.2.2. *Business model and value creation*

At a general level, a business model describes how to operate a company. Simplicity and practicality are often mentioned to describe BM. A BM represents the strategic positioning of the firm in a market [YIP 04] and defines how a firm creates and captures value for its stakeholders [CHE 07, CAS 09, CAS 10].

Amit and Zott [AMI 01] consider the BM as a unit of analysis that captures the value creation potential created by transactions between a focal firm and external stakeholders (partners, vendors and customers). The business model spans firm and industry boundaries. In addition, the two authors propose an activity system perspective for the design of business models based on two criteria: *design elements* (content, structure and governance) that describe the architecture of the activity system and *design themes* (novelty, lock-in, complementarities and efficiency) that describe the sources of value creation of the activity system.

	Traditional management accounting	Value chain analysis in the strategic framework
Focus	Internal	External
Perspective	Value added	Entire set of linked activities from suppliers to end-use customers
Cost driver single	A single fundamental cost driver pervades the literature – cost is a function of volume. Applied too often, only at the overall firm level.	Multiple cost drivers: Structural drivers (e.g. scale, scope, experience, technology and complexity); executional drivers (examples: participative management, total quality management, plant layout); each value activity has a set of unique cost drivers.
Cost containment philosophy	Cost reduction approached via responsibility centers or via product cost issues.	Cost containment is a function of the cost driver(s) regulating each value activity. Exploit linkages with suppliers Exploit linkages with customers "Spend to save"
Insights for strategic decisions	None are readily apparent. This is a major reason why the strategic consulting firms typically throw away the conventional reports as they begin their cost analysis.	Identify cost drivers at the individual activity level: develop cost/differentiation advantage either by controlling those drivers better than competitors or by reconfiguring the value chain. For each value activity, ask strategic questions pertaining to: make versus buy forward/backward integration quantifying and assessing "supplier power" and "buyer power"; exploiting linkages with suppliers and buyers.

Table 4.2. *Value chain analysis versus traditional management accounting (adapted from [SHA 92])*

If the concept of BM is not yet very well defined in academic literature, the analysis of the various definitions (Table 4.3) available reveals a number of recurring elements among which are the means of generating revenue, the resources and competencies required and the ways transactions between participants are organized. Based on these elements, the BM explains how the resources and competencies marshaled by a company allow it to develop a value proposition for its various client groups and how it consequently orders its internal value chain and value network.

Authors	Definitions
[AFU 00]	"A business model can be conceptualized as a system that is made up of components, linkages between the components, and dynamics" (p. 4).
[AMI 01]	"A business model depicts the content, structure and governance of transactions designed so as to create value through the exploitation of business opportunities" (p. 511).
[APP 01]	The author perceives a BM as a description of a complex business that enables the study of its structure, of the relationships among structural elements and of how it will respond to the real world.
[CAS 10]	"A business model is [...] a reflection of the firm realized strategy" (p. 195).
[CHE 02]	"In the most basic sense, a business model is a model of doing business by which a company can sustain itself that is, generate revenue (...). The essence of the idea is 'how you get paid', or 'how you make money' with a taxonomy of alternative mechanisms (...). The BM is "the heuristic logic that connects technical potential with the realization of economic value" (p. 529).
[EIS 02]	The BM is viewed as "a hypothesis about how a company will make money over the long term: what the company will sell, and to whom; how the company will collect revenue; what technologies it will employ; when it will rely on partners; and, following from the last two points, how its costs will 'scale' with growth" (p. xii).
[HAR 01]	The two authors point to the central role of the design of the business system, which is centered on "a particular set of understandings and interactions" (p. 494).
[HAM 02]	The author develops a complete approach to business models including: – the definition of the market scope: how to go to market? Competitors? – customer relationship management (CRM) and positive feedback effects: how to reach customers? – differentiation advantage, strategic assets and core competencies; – branding; – partner network.

[HAW 01]	A BM is described as the commercial relationship between a business enterprise and the products and/or services it provides in the market. He explains that it is a way of structuring various cost and revenue streams such that a business becomes viable, usually in the sense of being able to sustain itself on the basis of income it generates.
[HAB 12]	The author makes a distinction between a technology-driven disruptive business model innovation and a market-driven disruptive business model innovation. "We define an innovation where R&D experimentation precedes market opportunities and a business model development that will over time affect the incumbent firm's established market, as a *technology-driven disruptive business model innovation*. In contrast, a less sophisticated technological business model innovation that results from radical changes in the established value propositions to the existing customer [...], or altering the firm's role in the existing value chain or both [...], that will over time affect the established market, can be referred to as a *market-driven disruptive business model innovation*. Frequently, business model innovation emerges at a later stage when a once radical or disruptive technological innovation matures and competition through a new business model becomes critical" (p. 291).
[LIN 00]	Linder and Cantrell differentiate between three different types of models: the components of a business model, real operating business models and change models. They define a business model as an organization's core logic for creating value.
[MAG 02]	The author conceptualizes business models as "stories that explain how enterprises work. A good business model answers Peter Drucker's age old question: who is the customer? And what does the customer value? It also answers the fundamental questions every manager must ask: how do we make money in this business? What is the underlying economic logic that explains how we can deliver value to customers at an appropriate cost?" (p. 87).
[MEN 00]	The performance of the firm depends on the "architecture of the entire business network" (p. 519).
[MOR 05]	A BM is a "concise representation of how an interrelated set of decision variables in the areas of venture strategy, architecture, and economics are addressed to create sustainable competitive advantage in defined markets" (p. 727).
[OST 05]	"A business model is a conceptual tool that contains a set of elements and their relationships and allows expressing a company's logic of earning money. It is a description of the value a company offers to one or several segments of customers and the architecture of the firm and its network of partners for creating, marketing and delivering this value and relationship capital, in order to generate profitable and sustainable revenue streams" (p. 15).

[PET 01]	BMs are perceived as the logic of a business system for creating value. The authors specify that this is in opposition to a description of a complex social system itself with all its actors, relations and processes.
[RAP 01]	The author provides a comprehensive overview of the different views on business models. He identifies 29 different types of business models, ordered in nine categories.
[TAP 00]	"Business model refers to the core architecture of a firm, specifically how it deploys all relevant resources" (2001, p. 5).
[TEE 10]	"A business model articulates the logic, the data and other evidence that support a value proposition for the customer, and a viable structure of revenues and costs for the enterprise delivering that value" (2010, p. 179).
[TIM 98]	"An architecture for the product, service and information flows, including a description of the various business actors and their roles; a description of the potential benefits for the various business actors; and a description of the sources of revenues" (1998, p. 2).
[VEN 98]	"The business model is a coordinated plan to design strategy along all three vectors [customer interaction, asset configuration and knowledge leverage]" (p. 46).
[WEI 01]	An e-BM works "as a description of the roles and relationships among a firm's consumers, customers, allies and suppliers and it identifies the major flows of product, information, and money, as well as the major benefits to participants" (2000, p. 34).

Source: Developed by the author, based on analysis of the articles cited.

Table 4.3. *Business models' definitions*

The debate on BM definitions and main components refers also to normative approaches of BM, implying that a BM must take into consideration various aspects. Following a similar approach, Teece [TEE 10] states that:

> "business model design involves assessments with respect to determining: (1) the identity of market segments to be targeted; (2) the benefit the enterprise will deliver to the customer; (3) the technologies and features that are to be embedded in the product and service; (4) how the revenue and cost structure of a business is to be 'designed' (and, if necessary, 'redesigned') to meet customer needs; (5) the way in which technologies are to be assembled and offered to the customer; and (6) the mechanisms and manner by which value is to be captured, and

competitive advantage sustained. These issues are all interrelated. They lie at the core of the fundamental question asked by business strategists – which is how does one build a sustainable competitive advantage".

These approaches help managers to think and design their business models. In addition, business models will be increasingly digitized and "business intelligence" based models [DAV 06] should replace less efficient business models.

4.2.3. *Linking business models to other management tools (the balance scorecard)*

"If you can't measure it, you can't manage it" [KAP 92].

The balanced scorecard (BSC) is an effective strategic planning tool that gives managers a general overview of how well the organization is succeeding in meeting its mission and vision. The concept of the BSC was developed in the early 1990s by Kaplan and Norton [KAP 96]. It was originally conceived as an improved performance measurement system in order to determine if the organization is properly aligned and to improve shareholder value.

The BSC focuses on both financial and non-financial performance targets and outcomes. The BSC is a logical strategic framework organized across four key perspectives [KAP 00] leading to the identification of the critical drivers of success:

– financial perspective: increase value from new products and customers, increase customer value, improve cost structure, and improve asset utilization;

– customer perspective: customer value proposition;

– internal perspective: focus on processes that create new products and services, customer management processes, operations and logistics processes and regulatory and environmental processes;

– organizational learning and growth perspective: employee competencies, technology, corporate culture.

The BSC can be considered as a prescriptive framework that translates the organization's strategy into several perspectives, with a balance between short-term and longer term strategic goals, internal and external measures, performance results and the drivers of future results. Figge *et al.* [FIG 02] add that the BSC is a management tool that supports the successful implementation of corporate strategies in order to create more value.

Several authors try to provide an integrated framework for linking the BSC to BM (and particularly to e-BM) through identified values and strategies to facilitate strategic management activities. Yu [YU 05] proposes a BM-BSC strategic management approach. The BM-based BSC framework integrates value-based objectives regarding several perspectives (market, supply chain, customer, business structure and process) and various performance indicators: level of market competitiveness, market revenues, customer satisfaction level, return on asset, cash flow ratios, etc.

The digital transformation forces companies to reevaluate their value chain, business models and their accounting practices. It is the topic of section 4.3.

4.3. The digitization and the evolution of accounting practices: French perspectives

The adoption of digitization also means that companies start to think "digital-first" while they are developing new services and products and to transform their own organization, functions and processes. *What about accountancy?*

4.3.1. *From digitization to digital transformation*

Digital transformation is a polysemous buzz word. There are many dimensions of digital transformation sometimes confused with other terms such as digitization or digitalization.

Digital transformation is often considered as the next step of digitization. Digitization is the process of converting information from an analog to a digital format. The term "digitalization", closely related with computerization, was used first by Robert Wachal in 1971 who discussed the social implications of the "digitalization of society". In the Oxford English Dictionary, digitalization refers to "the adoption or increase in use of digital or computer technology by an organization, industry, country, etc.". The Gartner's IT glossary defines digitalization as "the use of digital technologies to change a business model and provide new revenue and value-producing opportunities; it is the process of moving to a digital business".

With regard to digital transformation, definitions are numerous, elaborated by both practitioners and academic scholars. Patel and McCarthy [PAT 00] have been among the first authors to address the digital transformation issue by highlighting three questions as follows: what is digital transformation? How does it change business? What are the challenges for B2B companies and B2C companies in making a digital transformation?

Table 4.4 presents an overview of definitions of digital transformation summarizing key focus words contained in the definitions.

Authors	Definition	Key features/words
[STO 04]	"Digital transformation can be understood as the changes that digital technology causes or influences in all aspects of human life" (p. 689).	General impact on business and society
[WES 11]	"The use of technology to radically improve performance or reach of enterprises" (p. 5). "Executives are digitally transforming three key areas of their enterprises: customer experience, operational processes and business models" (p. 17).	Performance
[HES 16]	"Digital transformation is concerned with the changes digital technologies can bring about in a company's business model, which result in changed products or organizational structures or in the automation of processes. […] (The) conceptual framework for formulating a digital transformation strategy identifies the four key dimensions of every digital transformation endeavor: *The use of technologies* reflects a firm's approach and capability to explore and exploit new digital technologies. *Changes in value creation* reflect the influence of digital transformation on a firm's value creation. *Structural changes* refer to the modifications in organizational structures, processes and skill sets that are necessary to cope with and exploit new technologies.	Strategy Changes in business models, organization, process, skills Value creation

	The *financial aspects* dimension relates to both a firm's need for action in response to a struggling core business as well as its ability to finance a digital transformation endeavor" (p. 124).	
[DOR 15]	Digital transformation is "creating value at the new frontiers of the business world, creating value in the processes that execute a vision of customer experiences and building foundational capabilities that support the entire structure" (p. 1).	Marketplace capabilities (proactive decision making; contextual interactivity; real-time automation; journey-focused innovation; agility;)

Table 4.4. *Digital transformation: a complex reality*

4.3.2. *Uber, uberization and digitization*

In recent years, in the context of digital transformation, a number of studies have been performed to gain further insight into the uberization phenomenon (based on digitization) notably through empirical publications. The term (neologism) of uberization has been coined by Maurice Lévy, CEO of Publicis Groupe to describe the disruption in particular of traditional business models by digital trends.

The particular attention given to uberization (even in non-specialized press) shows how its importance has grown year by year since the end of the 2000s [DAI 17]. The definitions given in the general and specialized press are generally broad or applied to a particular industry or activity (Table 4.5).

There are also several implications of uberization for accounting firms and practitioners (Table 4.5). Uberization could be seen as a threat as many basic accounting activities and daily work will be simplified and automated but it should also strengthen the strategic role of accountants in order to create more value for their clients.

> "Historically, management accountants have played a relatively indirect role in strategy determination-providing information seen as having strategic implications [BHI 09]".

Uberization of work – employment	"But of all the ways that Uber could change the world, the most far-reaching may be found closest at hand: your office. Uber, and more broadly the app-driven labor market it represents, is at the center of what could be a sea change in work, and in how people think about their jobs. You may not be contemplating becoming an Uber driver any time soon, but the Uberization of work may soon be coming to your chosen profession" [MAN 15]. "There has been a lot of debate about how online platforms have changed the nature of work. In some cases, on-demand companies have been harshly criticized for making employer like demands on workers but denying them basic benefits and protections" [BER 16].
Uberization of finance and banking	"We are on the verge of the Uberization of finance, which will bring multiple new opportunities but also a range of new risks. The ubiquitous ride-sharing company uses a simple device – the smartphone – to connect people who want rides with people who want to drive them. Uber is a high-tech middleman that is making the intermediaries of the past obsolete. The financial world is one of the most mediated industries on the planet, and that is precisely what is about to change. Uberization also means using vast amounts of data to make those connections feasible" [KAR 15].
Uberization of accounting technology and activities	"You have already seen a major push towards cloud tech, hosted software and integration of utility software with accounting "platforms". The "service delivery" has been Uberized to a large extent. Going are the days of clients waiting for paper documents to arrive from the accountant's office. (…) As key accounting technology platforms embrace open architecture, more and more of relevant data will start flowing automatically without much manual intervention. The time that you spend now in handling and making sense from such data and information will suddenly be available to you – and you would want to use that for more face-time with clients and prospects. As data and information increasingly manage itself, an accounting professional's role will distinctly change from that of "production" (of accounting information) to "insights leverage strategist" [PAT 15].

Table 4.5. *From Uber to uberization*

The uberization of the economy in France has gone through several phases in accordance with the evolution of regulatory requirements leading to an increasing degree of openness of several sectors. As explained by Bulow von Revel and Bolardi [BUL 16], the "Uber economy" targets both the regulated and non-regulated sectors:

> "Beyond non-regulated individuals attacking traditional regulated sectors (e.g. taxis, hotels...), Uber-like models increasingly rely on regulated new-comers competing with regulated traditional players (e.g. doctors, lawyers...)".

Several phases can be identified:

– first phase (*pioneer wave*): consumer-facing regulated sectors are attacked, by new players relying on non-regulated workers (Uber) or assets (Airbnb);

– second phase (*extension*): sectors with non-regulated professions attacked, mostly in the B2B space such as IT developers, designers, logistics workers, etc.;

– third phase (*proliferation*): regulated sectors attacked by new players relying on regulated workers, e.g. healthcare online services relying on board-certified doctors. The "uberization" takes the form of digitally empowered freelancers (lawyers, accountants, doctors, etc.) coupled with extra services and lower costs.

	Traditional sectors		
New suppliers		Non-regulated	Regulated
		Phase 2	Phase 1
	Non-regulated	Car rental House cleaning Freelance Education Consulting Logistics	Hotels* Taxis
			Phase 3
	Regulated		Certified accounting Healthcare Legal

*Hotel activities are regulated per se, although their workers are not.

Table 4.6. *The "Uber economy" in France (adapted from [BUL 16])*

4.3.3. *What digital transformation of accounting activities?*

As we have shown previously, the digital transformation has a considerable impact at the external level especially on the different market players (traditional firms and new entrants). However, it also has dramatic effects within the company itself, its functions, activities and processes. This vision is not new. Several scholars wrote about it in the early 2000s:

> "There is mounting evidence that the deployment of digital technologies by organizations not only affects the economics of operational and managerial processes but also mobilizes extensive social and organizational effects. Digitization impacts the form, substance, and provenance of internal accounting information with attendant consequences on the behaviour and actions of organizational participants and on the functioning of enterprises more widely. Knowledge about the influence of the deployment of digital technologies on management accounting thinking, processes, and practices is starting to take shape" [BHI 03].

Several elements have to be considered when we analyze the evolution of accountancy practices in a context of digital transformation. At the technological level, [ACC 13] has identified 10 key technologies (mobility, cloud, social collaboration, digital service delivery, big data, payment systems, cyber security, robotics, augmented and virtual reality, artificial intelligence). All of them should have a great impact on accountants' practices in the near future. Accountants are already using digital services to provide resources and access resources. But new patterns will emerge because of the adoption of technologies such as augmented and virtual reality. The example of the general use of Second Life by accountants has been given [ACC 13] "to recruit trainees, attract clients and develop new lines of business, and [for] holding meetings in online role-playing games such as World of Warcraft".

Today, one of the most visible changes has been the simplification of invoice (number)/purchase (order). Invoice processing is highly automated because of machine learning in order to reduce processing cost. More generally, the main objectives are to improve the transparency of invoice documents to achieve better cash management and higher levels of payment accuracy.

In France, several companies have already launched the automation of processes associated with the dematerialization of accounting records (Box 4.1).

Orange France Telecom is a French telecommunications company. It employs nearly 172,000 people, including 105,000 in France, and serves nearly 226 million customers worldwide.

Key figures for Accounts Payable:

Approximately 1.2 million purchase invoices processed of which 450,000 are processed manually.

750,000 EDI invoices processed using the same workflows.

Cost of processing an invoice (non-EDI) after automation: 7 € instead of 12 €.

Reduced staff of 37 full-time equivalent for manual processing of invoices.

In 2006, France Telecom Group launched a major project to transform its finance department and enable "sales to cash," "procure to pay" and accounting process optimization. For this, the group decided to industrialize business processes through the implementation of an accounting shared service center built on a repository unified with the main management repository.

The foundation of this project is the implementation of a single ERP for all subsidiaries.

The CSP (shared services center dedicated to accounting), part of the finance information service of France Telecom Group, is an entity with over 750 staff managing accounting for the French subsidiaries. It is organized into eight sites across the country, each specializing in a specific process (amortizations, suppliers, business clients, public clients, etc.). It provides accounting operations for 30 subsidiaries with the two most important being Orange and France Telecom. The CSP manages the transactional activities in the ERP but also includes activities with high added value, such as closing accounts, which requires knowledge of IFRS standards or taxation. It requires financial information as soon as possible, to present management with a D+5 EBITDA and optimal visibility at any time. It uses a single source of reliable information.

Source: http://www.kofax.com/~/media/Files/Kofax/Case-Studies/cs-orange-france-telecom-en.pdf

Box 4.1.

4.3.4. *From big data to advisory services*

In today's dynamic business environment, one of the most crucial issues of digital transformation is closely related to the ways to manage the increasing volume and complexity of the data to be analyzed and audited. Big data is often described using five Vs: Volume, Velocity, Variety, Veracity and Value. The "big data" generic term refers particularly to intelligent predictive tools to anticipate customers' future needs.

As most firms are focusing on agility, responsiveness and customer centricity in order to remain competitive, technology is considered as a cost-effective way to essentially outsource basic accounting business functions. Accountants will gradually spend less time running day-to-day activities and will shift to the development of an advisory service (based partly on real-time reporting technology). In order to provide better advice to clients and to accelerate company performance (via KPIs), they will use more and more data analytics.

Big data do not target only firms. In France, at a macroeconomic level, public authorities have also developed various electronic databases (containing economic, fiscal and social information). As Arraou [ARR 15] has highlighted, major efforts have been made to get a general overview of very small enterprises (VSE) small and medium enterprises (SMEs) and to analyze the impact of public policies on them:

> "By telefiling their clients' fiscal and social data for several years now, certified public accountants contribute to building up one of the largest VSE/SME databases in existence. The Order of Certified Public Accountants is now giving new life to those documents, taking anonymised data and 'making it talk', much in the same way as certified public accountants get their clients' figures to 'talk'. By pooling resources and material, the Order is helping improve economic analysis in our country and shed light on the impact of legislative measures on the activities of French VSE/SMEs".

4.4. Conclusion

Digital transformation, buzzword or not, is vital to the success of today's business. The digital revolution is disrupting entire industries leading to the reconfiguration of the value chain and the launch of renewed revenue models. This digitization could lead also to a massive erosion of competitive advantages such as location and access to knowledge and information (big data).

Accountancy is also in need of digital transformation. Most of the firms are gradually switching their accounting and IT systems in order to improve their performance. In addition, today's organizations are becoming increasingly data driven and if accountants can develop a core competency in big data analytics, they will be able to efficiently support their clients.

Part 3

The Notion of Value Creation in the Context of Japanese SMEs

5

Value Creation from Voluntary Disclosure by Small- and Medium-Sized Entities

5.1. Introduction

In recent times, there has been increasing attention on the accounting practices of small- and medium-sized entities (SMEs) across the world regarding growing trends to reduce the cost and complexity of their financial reporting. This focus is primarily being driven by accounting standard setters worldwide (e.g. [PRI 13] for the United States; FIN 15] for UK; [INT 09]). For example, the International Accounting Standards Board issued a separate set of International Financial Reporting Standards for SMEs [INT 09]. With regard to this standard, extant literature has actively discussed the cost and value of adoption, especially among European companies [CEU 16, EIE 13a, LIT 12, NOB 10, EIE 09], because SMEs in these countries have discretion in the preparation of their financial statements. They may choose either IFRS for SMEs or other local Generally Accepted Accounting Standards (GAAP) for SMEs. In the US, [HOP 17] addressed SMEs' voluntary disclosure policy of preparing financial statement in accordance with fully fledged local GAAP when they have alternative standards that allow exceptions for private company reporting. These prior studies focused on whether or not voluntary financial disclosure by SMEs helps to reduce the cost from information asymmetry and create potential firm's values afforded for SMEs when preparing financial statements. However empirical evidence on motivations for adopting voluntarily disclosure schemes remains inconsistent in the accounting literature [EIE 13a, EIE 09].

In Japan, there are similar arguments regarding the choice of voluntary disclosure for SMEs (e.g. [KUS 15, MAN 12]). The prominent part of this argument is that there are multiple alternatives of financial disclosure schemes to choose from. For instance, there are two accounting standards especially designed for SMEs. These are the Accounting Standard Board of Japan Guidelines for SMEs (ASBJ Guidelines) and General Accounting Standards for SME (General standards). The SMEs that are not listed and not categorized as large companies are able to apply either of these standards as the GAAP for SMEs [KAW 14]. It is thought that ASBJ Guidelines are subject to IFRS framework, while the General Standards allow firms to avoid the impact of IFRS [KAW 12]. Nevertheless, both are usually applied interchangeably. In addition to these two options, SMEs are also allowed to apply accounting standards (or not restricted from doing so) for large companies that include the Japanese General Accepted Accounting Principles (J-GAAP) and International Financial Reporting Standards (IFRS) by their own choice. Conversely, SMEs may ignore all the above disclosure standards because they are not mandatory, thus they can choose to comply with tax accounting rules used when lodging tax returns. This is because financial reporting of SME is not regulated by the equity market so firms have wide discretion in preparation of financial statements and can voluntarily choose from a number of accounting standards.

Despite this extant opportunity to choose from multiple options for voluntary disclosure, little is known about Japanese SMEs' financial reporting practices, particularly in relation to factors SMEs consider when determining their production and use of financial statements. Previous accounting literature has extensively investigated the effect of SMEs' disclosure policy choice but these studies simply statistically examined the effect of specific proxies such as accrual quality [HOP 16, FUJ 15, HOP 13, ALL 09], stock liquidity [LAR 14] and various corporate characteristics [TOR 16, AGY 16, HOP 13]. No study to date has been undertaken to explore this argument by using more holistic perspectives.

Given this background, the following three chapters were developed to comprehensively explore the factors that determine the choice of financial accounting standards for implementing voluntary disclosure of financial reporting among SMEs in Japan, and also to investigate the associations between SMEs' disclosure choice and their value creation activities. First, the present chapter briefly describes the summary of the SME accounting scheme in Japan, theoretical foundations of the voluntary disclosure choice and literature reviews that consist of the premises for two empirical tests administrated in the following two chapters. Following this chapter, Chapter 6 performs a deductive test to statistically explore factors to determine the voluntary choice of accounting standards. Then, Chapter 7 conducts an inductive exploratory test to achieve the same research purpose from another methodological point of view.

5.2. SME accounting scheme in Japan

SMEs in Japan have a wide discretion in the preparation of financial statements from several alternative standards. These are as follows.

5.2.1. *Accounting Standard Board of Japan Guidelines for SMEs (ASBJ Guidelines)*

The ASBJ Guidelines were released in 2005 as the accounting standards for SMEs to describe accounting treatments and explanatory notes when preparing financial statements. Compliance is not mandatory, and only a small number of SMEs use these standards [KAW 14]. ASBJ Guidelines are regarded as the set of rules simplified from large firms' accounting standards to adjust for specific SMEs situations [KAW 12]. One distinguished trait of these Guidelines is to incorporate the IFRS framework.

5.2.2. *General Accounting Standards for SMEs (General standards)*

The General standards were issued in 2012. Similar to ASBJ Guidelines, these are not a mandatory standard. General standards are thought of as more stand-alone standards that reflect specific features of SMEs in the standard setting process and avoid influences from large entities' accounting schemes [KAW 14, KAW 12]. Further, General standards allow SMEs to completely avoid the impact of IFRS and even permit the application of several accounting rules from the Corporate Tax Act [KAW 12].

5.2.3. *Accounting Rules of Corporate Tax Act (Tax Accounting Standard)*

While ASBJ Guidelines and General Standards are voluntary financial disclosure schemes, the Tax Accounting Standard is a mandatory rule that all registered companies must follow when preparing and reporting their annual income tax returns. Many prior studies point out that in Japan there has been a long tradition of a close connection between financial and tax accounting through the so-called "principle of congruency" (e.g. [KAW 16, FUJ 15]). With this principle, SMEs managers often choose the Tax Accounting Standard when preparing financial statements simply to minimize the cost of tax return adjustments [FUJ 15].

5.2.4. *The International Financial Reporting Standards for SMEs (IFRS for SMEs)*

The IFRS for SMEs was originally developed by the International Accounting Standards Board (IASB) in order to provide SMEs internationally comparable financial accounting information. In Japan, the Business Accounting Council, which is the advisory board of the accounting standard setter in Japan, issued an official report titled "*An Interim Policy Relating to International Accounting Standard*", suggesting that SMEs' accounting practice should not be influenced by IFRS [BUS 13]. As a result, IFRS for SMEs are applied very infrequently by Japanese SMEs. However, in theory, IFRS for SMEs can be applied by any Japanese SME in order to prepare their financial statements.

5.2.5. *Japanese Generally Accepted Accounting Principles (J-GAAP)*

Japanese SMEs are not mandated to apply J-GAAP. For example, SMEs that intend to make an Initial Public Offering in the near future may choose this standard in advance in order to make the transition to a public company easier.

5.3. Theoretical foundations

The rationale for voluntary financial accounting disclosure research has often been explained by three theories: agency theory [HEA 01, JEN 76], signaling theory [SPE 73, MOR 87] and proprietary cost theory [VER 83]. Prior studies have often integrated these three theories into their theoretical framework to investigate potential determinants of voluntary disclosure (e.g. [PRE 04, DEP 00, INC 97]). In the present research, several adjustments will be made and reflected in the framework development since here the research setting is based on SMEs where there are many substantial differences compared to large public companies.

First, agency theory focuses on the conflict of interests and information asymmetry between owners and managers, where managers are expected to disclose more information to reduce agency costs [JEN 76]. In the SME setting, where owners and managers have little separation, this theory deals with the agency conflict more specifically between insiders (SME managers) and outside shareholders or lenders, who would be provided information about performance achieved by managers [LIT 12, PAT 07].

Second, signaling theory postulates that voluntary disclosure may be considered the direct signal of quality within the company, and therefore may be used to reduce the risk of adverse selection [MOR 87]. It is thought that information asymmetries tend to be greater in SMEs, because insiders have better information on their firms and company-specific information is not required to be disclosed in public. This may then influence a lender's decision to provide finance to SMEs [BUT 07]. Given the specific setting of SMEs, signaling theory indicates that one way for SMEs to reduce information asymmetries is through providing voluntary disclosure based on accrual accounting. This is because this sophisticated accounting method is more costly, and signals the high quality of the firm's accounting that then increases loan approval and possibly reduced interest rates for them [CAS 15, CON 11].

Lastly, proprietary costs theory must be taken into consideration when evaluating voluntary adoption of an accounting policy in this research. This theory states that companies limit voluntary disclosure of relevant information to the financial market because disclosure-related costs are harmful for the reporting company [LIT 12, PRE 04]. For example, SMEs may incur relatively higher costs for complying with financial reporting requirements than larger companies because they do not enjoy the accounting-specific economy of scale effects of large firms [EIE 09, EVA 05, COP 06]. Furthermore, it is believed that opportunity costs resulting from disclosure might be higher and not efficient for small firms, since small firms often have a lower volume of business [EIE 09].

5.4. Literature review

Studies on voluntary choice of comparable global financial reporting for SMEs first discussed the application issue for the IFRS for SMEs [KAY 15, KIL 14, UYA 13, ALB 13b, ALB 12, LIT 12, BUN 12, ATI 10, EIE 09]. Among these studies, Kiliç *et al.*.[KIL 14], Uyar and Güngörmüs [UYA 13] and Bunea *et al.* [BUN 12], for instance, examined the perceptions and knowledge of accounting professionals pertaining to the content and implementation of IFRS for SMEs. These studies addressed IFRS for SMEs in terms of awareness, technical knowledge, advantages and obstacles of implementing this standard. However, some previous studies pointed out how SMEs deal with the IFRS for SMEs has been little investigated especially among developed countries (e.g. [QUA 12]). In this regard, Uyar and Güngörmüs [UYA 13] also suggests further studies on this topic are needed in developed countries to provide different feedbacks and reflections for future directions in this research field.

The primary reason why less research for IFRS for SMEs exists in developed countries was well explained by Kaya and Koch [KAY 15]. This study used 128 countries' data to investigate associations between jurisdiction systems and adoptability of IFRS for SMEs. The authors confirmed that IFRS for SMEs was mainly applied in developing countries and concluded that lower demand for IFRS for SMEs in developed countries was because of their strong ability to develop local accounting standards for SMEs by themselves.

Eventually in the United Kingdom, the Financial Reporting Council (FRC) issued the Financial Reporting Standards (FRS) 102 in 2015, which was mainly based on IFRS for SMEs, but modified significantly in order to be consistent with company law of UK [KAY 15]. In the United States, the American Institute of Certified Public Accountants (AICPA) also developed its Own Comprehensive Basis of Accounting (OCBOA) financial reporting framework for SMEs in May 2012, which blended traditional accounting methods with local income tax requirements [SUN 13]. Although private companies in the United States can choose to voluntarily apply IFRS for SMEs, most of them have selected to prepare their financial statements on OCBOA or selected to report US GAAP exceptions [SUN 13]. In all European Union (EU) countries, adoption of IFRS for SMEs inevitably requires a legislative change in the law of each country to adjust tax and company laws to a new set of accounting standards [KAY 15]. However, the application of IFRS for SMEs implied breaking the traditional bond between financial statements and the income tax return [EPI 10], which made several countries such as France and Germany opposed to this standard [ALB 12].

Irrespective of IFRS for SMEs or local SMEs accounting standards, SMEs take into considerations of various criteria to make their choice for applying global comparable financial reporting standard. Previous studies attempted to identify determinant factors for SMEs' decisions to adapt global standards. Among these studies, many studies have examined the role of a firm's size as a relevant proxy to determine application of either local or global standards [EIE 13a, ALB 13b, AND 12, BUN 12, LIT 12, QUA 12, EIE 09]. Eierle and Haller [EIE 09], for instance, conducted an empirical study using German SMEs to explore size as an effective factor for accounting differentiation. As the result, the authors demonstrated the entity's economic size had a significant impact on relevance of international activities and involvement of owners in management. Other empirical research in the UK by [AND 12] also expected that higher level of information disclosure could reduce various types of capital costs. Their multivariate logistic regression revealed that firm size predicted significant positive impact on UK unlisted firms' choice in voluntary adoption of IFRS.

In contrast, [EIE 13a] considered a firm's size as the control variables (total assets and the number of employers) of their main investigation regarding criteria for implementing global comparable financial reporting, but they found no significant evidence in a firm's size as the effective proxy. This result was inconsistent with [EIE 09], although analysis models and techniques applied were different between two studies. Followed by [EIE 09], Carfang [CAR 10] theoretically criticized that size criteria was not the determining factor in establishing the scope of the IFRS for SMEs application because any quantified size criterion is not suitable with a view of IFRS's principled-based approach. Accordingly, the literature presented mixed results regarding the correlation between the size of the business and the level of international standards adoption.

In the prior literature, ownership structure was also considered to be another key driver to determine the voluntarily choice of global accounting standards. In theory, the voluntary choice made by firms to adapt a specific accounting policy or standard such as IFRS for SMEs can be considered as a mechanism for the reduction of agency cost resulting from the separation of ownership and management of the firm [LIT 12]. Given this theory, [EIE 13a] empirically confirmed that a higher proportion of owners involved in the management would decrease firm's need to provide internationally comparable accounting information for owners. This finding was also consistent with Bassemir [BAS 12] and Francis, Khurana, Martin *et al.* [FRA 08] who reported an impact of the ownership structures on SMEs' decision to voluntary disclosure of the global comparable financial information, although these studies dealt with full IFRS adaptation issue. On the other hand, a UK study of André, Walton and Yang [AND 12] failed to find the significant association of the voluntary use of full IFRS with the separation between owners and managers. This study inconsistently concluded that ownership structure did not affect the decision to choose full IFRS by UK unlisted firms. Thus, results of this explosion of interest are mixed and difficult to synthesize into a meaningful whole.

Prior literature also pointed out that SMEs would be motivated to choose global comparable standards in order to make the transition easier to apply full IFRS particularly among growing SMEs. Much evidence of this phenomenon has been observed in the emerging economies ([KIL 14] in Romania; [UYA 13] in Turkey; [MUL10] in the Czech Republic). Similar to this transition purpose, intent to have an IPO [HIE 13, BAS 12] was also found to be another crucial determinant for SMEs' decisions to adapt global standard voluntarily.

Another research line for examining factors to drive voluntary choice of global comparable reporting standards is to analyze the cost-benefit relationship [EIE 13b]. In line with this research stream, some prior studies admit that SMEs' benefits or demands of adapting global comparable reporting have a strong correlation with the possibility of their actual application of the standard, but these benefits or demands vary extremely between the SMEs (e.g. [EIE 13a, AND 12, BAS 12]) and depend on national aspects (e.g. [WEH 14]).

For example, many studies dealing with voluntary application of IFRS for SMEs in emerging countries have stressed that to voluntarily apply global accounting standards improves qualities of the financial reporting (e.g. [KIL 14] for Turkey; [BUN 12] for Romania). These researchers found that accounting professions in these countries viewed IFRS for SMEs would add values such as comparability, reliability, transparency and understandability to their financial reporting. With regard to these benefits, Albu, Albu and Gîrbină [ALB 13b] contended that such benefits specifically anticipated in emerging economics would differ from those expected in more developed countries. This prior study examined stakeholders' perceptions in four Central and Eastern European emerging economies and concluded that emerging countries expect to yield benefits that developed countries did not seek, such as attractions of foreign investment and enhancement of economic development [ALB 13b, BRO11].

In a similar line of research, Uyar and Güngörmüs [UYA 13] and Kiliç *et al.* [KIL 14] collected their data in Turkey to examine accounting professions' view toward IFRS for SMEs. Both studies reported that increase in efficiency of cross-border activities was evaluated higher as the possible benefit of implementation of IFRS for SMEs. Albu and Albu [ALB 12] also investigate implementation of IFRS for SMEs in Romania. Using interviews over various stakeholders surrounding SMEs, they also demonstrated that the interviewers agreed the positive benefit of IFRS for SMEs' implementation enhancing opportunities for doing business with foreign entities. However, Uyar and Güngörmüs [UYA 13] emphasized that other aspects of ease in reaching global capital and increase in foreign direct investment in Turkey did not receive much support from respondents. Regarding voluntary adaptation of global standards in the emerging countries, Kaya and Koch [KAY 15] summarized that their switch to IFRS for SMEs could facilitate the contracting environment to attract loans from international organizations such as the World Bank and the IMF. But it was not certain whether this positive reputation effect from the adaption of global standards would be reflected to enhance more active foreign trading of SMEs.

In contrast to research in the developing and emerging countries, there are several studies empirically addressing the SMEs' voluntary disclosure of the global accounting information in the developed countries. Eierle and Helduser [EIE 13a], for instance, investigated the need for unlisted SMEs in Germany to provide global comparable financial accounting information. With a sample of 322 SMEs, the authors found that unlisted SMEs manifesting a need to provide global relevant accounting information were more often financed by international capital providers including foreign investors and lenders. Andre *et al.* [AND 12] also explored UK unlisted firms and the number of their subsidiaries that are supposed to be financed with local external capital and regulated by local market. Although this research addressed the voluntary choice of the full IFRS issue, it supports that firm's internationalization acts as an important driver in this accounting choice. In addition, Bassemir [BAS 12] conducted a similar study with [EIE 12] by using a sample of German private firms, and found that higher proportion of foreign exports significantly raised a private firm's probability of full IFRS adoption. However, Eierle and Helduser [EIE 13a] failed to find significant results in the predictive power of foreign exports and imports on the voluntary adoption of global SMEs accounting. Accordingly, the internationality is still controversial as an effective factor for determining the voluntary choice of global accounting for SMEs.

5.5. Association between SMEs' voluntary disclosure and value creation

The argument of voluntary disclosure is also related to how to define the corporate value creation. In the accounting literature, some previous studies regard firm's value from voluntary disclosure as the conventional financial benefit (e.g. [HEA 01, HAN 00]), whereas other studies extended the concepts of the value, which includes even non-financial benefits [CIA 13, STE 10]. The former line of the studies was interested prominently in the relationship between voluntary disclosure and stock prices as well as the company's cost of raising capital. For instance, Healy *et al.* [HEA 99] found that firms with significant improvements in voluntary disclosure experience increased stock performance and capital market intermediation. Similarly, Leuz and Verrecchia [LEU 00] discovered that voluntary reporting by using International Accounting Standards or US GAAP was significantly associated with lower bid-ask spreads and higher share turnover.

In contrast, the latter research line addressed the effect of other non-financial and intangible aspects, including reputation, sustainability, market conditions and pressures from corporate governing bodies. Prior studies report that such aspects would drive specific voluntary disclosure practices of companies (e.g. [BOE 07, AHM 99]). Among them, Boesso and Kumar [BOE 07] attempted to investigate what factors drove the voluntary disclosure practices within Italian and US firms. They found that variety of factors that related to not only investors' information needs but also sustainable value creation for all stakeholders affecting both the volume and the quality of voluntary disclosure. The corporate values generated from non-financial aspects are often explained by legitimacy theory, which provides theoretical motivation for social disclosure by the companies. This theory assumes that there is a social contract between a company and the society that requires the company to be responsive to the environment in which it operates [DEE 00]. In such a theoretical setting, the role of the accountant becomes to provide information about sustainable development activities [DEV 01, VAN 09]. With regard to the SMEs, Mistry, Sharna and Low [MIS 14], however, examined the perception of management accountants of SMEs' sustainable development and discovered that SMEs did not have dedicated roles in accounting for sustainable development because of their lack of resources. Eventually several studies have shown that businesses were linked with their natural environment, of which appropriate treatments would improve corporate performance and contribute to competitive advantage (e.g. [CIA 13]). Steenkamp and Kashyap [STE 10] also provided the empirical evidence of SME managers' perception about the importance of and the contribution of intangible assets to their companies and found that they perceived understanding, identifying and managing intangible assets were important value drivers for their companies. These studies, however, failed to present solid evidence as to what is exactly meant by the SMEs' corporate value generated from voluntary disclosure.

5.6. Summary

The purpose of this chapter was to describe the summary of the SME accounting scheme in Japan, theoretical foundations of the voluntary disclosure choice and literature reviews. Little accounting literature was found to address factors affecting the voluntary disclosure of financial reporting among SMEs in Japan and its influence toward their value creation activities and decision making. The components demonstrated in this chapter become seminal premises for two empirical tests discussed in Chapters 6 and 7.

6

SMEs' Value from Voluntary Disclosure: Deductive Quantitative Approach

6.1. Introduction

As seen in Chapter 5, the existing literature has attempted to identify the crucial factors to determine the voluntary adaptation of global comparable financial reporting for SMEs, but the findings have not reached a consensus. Furthermore, despite increasing research activities related to this topic overseas, no empirical study of this field has been undertaken in Japan. Japanese SMEs are confronting the controversial choice of voluntary adaptation to either Accounting Standard Board of Japan Guidelines (ASBJ Guidelines) that basically reflect the impact of IFRS, or General Standards that allows SMEs to avoid any IFRS influence and stay with domestic tax purpose accounting [KAW 12]. This choice of SME accounting standards highly relies on Certified Public Tax Accountants (CPTAs) who engage with SMEs as tax and management consultants. Regarding this association between CPTAs and SMEs, Teikoku Data Bank [TEI 13] investigated SME accounting in Japan by using a questionnaire-based survey among SMEs executive managers and revealed that 65.8% of managers replied that they did not know about their companies' accounting activities because they were relying on their hired accounting professionals. Research exploring this issue through perceptions of professional accountants in Japan is important, but relatively sparse.

Among the few studies on ASBJ Guidelines, Murata [MUR 10], for example, conducted a questionnaire-based survey to examine CPTAs' various perceptions toward ASBJ Guidelines and discovered that 40.3% of participants believed that the primary reason for choosing ASBJ Guidelines was to make financing activities easier. On the contrary, Murata [MUR 10] also found that the traditional bond between financial and tax accounting (principle of congruence) in Japanese SMEs was a reason to avoid applying the ASBJ Guidelines. These findings may imply that CPTAs' choice of SME accounting standards would often be diverted for the earings management purpose by using the tax accounting scheme, which also allowed SMEs to apply by the local GAAP of the General Standards.

The literature in Japan remains inconclusive about the factors affecting the choice of SME accounting standards. Kushibe [KUS 14] and Mandai [MAN 12], for instance, proposed that only relatively large SMEs will be allowed to apply ASBJ Guidelines, while other smaller SMEs could choose General Standards, although some overseas research showed no statistical correlations between a firm's size and their voluntarily choice of global accounting standards (e.g. [EIE 13, CAR 10]). Compared with these studies, Uenishi [UEN 12b] and Uezu [UEZ 14] indicate that a firm's choice of SME accounting standard would be subjected to the firm's various characteristic including their intention of IPO, financing purpose and accounting information demands. Accordingly, the findings in the previous studies are still contradictory and inconclusive.

Given the conflict between global and domestic accounting standards for SMEs in Japan, out research attempted to address this accounting choice issue from Japanese CPTAs' perspective. First, we explored CPTAs' opinions about ASBJ Guidelines as to whether or not they agreed with the need for this standard. Although Murata [MUR 10] examined perceptions of CPTAs about ASBJ Guidelines, no study has been conducted on this topic after the publication of the General Standards in 2012. Second, this prior study tried to identify the determinant factors to drive SMEs' voluntary adaptation of global comparable financial reporting standards (ASBJ Guideline). Accordingly, the following two research questions (*RQ*) were formulated in this present chapter to address our research gaps presented in the Chapter 5.

RQ 1: Do the majority of CPTAs in Japan agree with applying global comparable accounting standards (ASBJ Guidelines) or not?

RQ 2: What are perceived determinants to explain SMEs' choice to voluntarily adaptation of global comparable accounting standards (ASBJ Guideline) among CPTAs in Japan?

As seen in the literature overseas, benefit and cost analysis from adaptation of global comparable reporting by SMEs is of great interest. However, few studies to date have been undertaken to focus on SMEs in Japan. Among the research in Japan, Teikoku Data Bank [TEI 13], which was consigned by the Small and Medium Enterprise Agency (SMEA) in Japan, published an official report empirically investigating various situations about SME accounting standards. Along with Uyar and Güngörmüs [UYA 13] in Turkey, Teikoku Data Bank [TEI 13] asked SME managers who were familiar with accounting to provide perceived values of applying SME accounting standards for their companies (e.g. reliability, transparency, domestic finance, effective management). Unfortunately, this study failed to ask perceptions separately about ASBJ Guidelines and General Standards.

Given the literature, this chapter focuses on values of adapting SME accounting standards. In this chapter, both values of applying ASBJ Guidelines and General Standards are addressed and compared, rather than observed costs of choosing global accounting standards for SMEs. This is because the values of both standards stand for the other side of the same coin. Comparative analysis of both values will provide important insights regarding the associations between two standards.

Our research also attempts to identify whether the values generated from SMEs' voluntary disclosure would be of conventional financial benefit or non-financial benefit. It is true that most previous studies have analyzed the role voluntary disclosure plays in creating economic power and value (e.g. [HEA 01, HAN 00]), while other research supports that non-financial factors such as sustainable development and long-term value creation would be the factors driving the voluntary disclosure practices of companies [AHM 99, MAR 91]. The values of voluntary application to SME accounting standards will be measured according to CPTAs' perceptions. For this purpose, the following three research questions were developed.

RQ3: What are the perceived values of applying global comparable accounting standards for SMEs (ASBJ Guidelines) among CPTAs in Japan?

RQ4: What are the perceived values of applying local GAAP for SMEs (General Standards) among CPTAs in Japan?

RQ5: Are there any differences in CPTAs' perceived value between applying ASBJ Guidelines and General Standard?

6.2. Research design

6.2.1. *Participants*

The data were collected from accounting professionals who are certified and registered as CPTAS in Japan. The sample was identified by the web-based database (https://www.zeirishikensaku.jp) operated by the Japan Federation of Certified Public Tax Accountants' Association (JFCPTAA). In this database, the JFCPTAA members voluntarily disclose their professional information including name, affiliation, contact details and e-mail address. For research purposes, the present study identified the CPTAs who engage with corporation tax practices at the small- or medium-sized individual accounting firms. The participants of this survey were either employers or employees of the accounting firms. A total of 471 CPTAs were sent an e-mail on June 22, 2015. The deadline date of response was August 22. The e-mail sent from us provided a direct link to the web-based survey system (www.surveymonkey.com), which enabled participants to access to the questionnaire developed by the authors. As the result, 69 responses were collected through this web system. Among them, 14 responses were excluded from the final data set due to incompletion of the questionnaire. The effective response was 55, giving a response rate of 11.67%.

6.2.2. *Questionnaire development*

The questionnaire consisted of three sections. The first section included survey items that asked for the participants' perceptions regarding global accounting standards for SMEs. The participants were asked to respond whether or not they agree to apply the ASBJ Guidelines incorporating IFRS-based global comparable accounting standards. The questionnaire also asked them the reasons for their responses as to why they agreed and what they thought was the best use of ASBJ Guidelines in order to coexist with General Standards.

For research purposes, this study codified descriptive information from participants regarding the reasons and the ways of coexisting with ASBJ Guidelines into more concrete criteria that sampled CPTAs perceived to use for SMEs' choice between ASBJ Guidelines and General Standards. To codify the information, a coding schedule was designed to capture descriptive information of the data with the key criteria appearing in several prior studies [KIL 14, EIE 13, UYA 13, ALB 13a, AND 12, LIT 12, BAS 12, YAN 12]. The inter-rater reliability in the coding of data between researcher and research assistant was checked in this study. No inconsistencies between two coders were found. These results confirm high levels of reliability.

The second section of the questionnaire contained survey items to ask about the participants' perceptions toward values of both ASBJ Guidelines and General Standards application, respectively. In accordance with several previous studies [KIL 14, UYA 13, TEI 13], 10 items of perceived values (comparability, reliability, transition purpose, overseas finance, domestic finance, direct investment, efficient management, effectiveness for M&A, effectiveness of auditing, effectiveness of taxation) were prepared to ask for each two accounting standard. The five-point Likert scale method, which goes from one for strongly disagree to five for strongly agree, was applied to measure these variables.

The final section comprised several questions about the participants' descriptive statistics, including their age (AGE), gender (GENDER), duration of work experience (WORK EXPERIENCE), status of employment (POSITION), number of clients (CLIENT), awareness of SMEs accounting standards (AWARE) and implementing rate of standards apply for their clients (IMPLEMENT).

6.2.3. *Descriptive statistics*

Table 6.1 provides the participants' descriptive statistics. This table reports that the mean age for participating tax accountants was 51.41 years old. In terms of gender, 52 subjects (94.5%) were male and only two subjects (3.6%) were female. It is also demonstrated that the majority of participants were working as employees (46 persons; 83.6%), while nine subjects (16.4%) were running their own accounting firms. As for their work experience, 33 participants (60%) had 10–30 years of work experiences. 19 subjects (35.2%) had more than or equal to 100 and less than 300 SMEs as their regular clients.

Preliminary tests (t-test, MWU-test and chi-square test) were performed to check the homogeneity between the CPTAs who agreed with ASBJ

Guidelines and those who did not. Significant differences in these descriptive variables between these two CPTA groups would cause serious distortions to the primary analysis of this study. However, the outcomes of these tests did not report any significant differences in AGE (t = 0.284, P = 0.778; z = 0.200, P = 0.841), GENDER (χ^2 = 0.260, P = 0.610), WORK EXPERIENCE (χ^2 = 1.893, P = 0.755), POSITION (χ^2 = 0.723, P = 0.395) and CLIENT (χ^2 = 4.114, P = 0.391) and assured the homogeneity of the two groups.

Table 6.2 presents the participants' awareness of the two different SME accounting standards and to what extent they actually applied each standard for their clients. In panel A, it is reported that nearly 95% of subjects in this study had a detailed knowledge of both ASBJ Guidelines and General Standards. These data are consistent with the latest statistics from Fujibayashi *et al.* [FUJ 15] which report empirically investigated awareness of SME accounting standards among 723 tax accountants in Japan and found 100% of their subjects knew about General Standards. This report did not examine the awareness of ASBJ Guidelines, but our research discovered that CPTAs in Japan were fully aware of ASBJ Guidelines as well.

According to panel B in Table 6.2, General Standards were applied for 81.8% of participants' clients, while ASBJ Guidelines were applied only for 40.7% of clients. Fujibayashi *et al.* [FUJ 15] also asked a similar question about compliance rate to both ASBJ Guidelines and General Standards among their participated professions. Although the figures reported were not exactly similar due to the different question formats, the tendency that application rate of General Standards was dramatically higher (93.2% including both full and partial compliance with standards) than that of ASBJ Guidelines (66.9%) was evident in the prior report.

AGE	Total	CPTA who agree with ASBJ Guidelines	CPTA who disagree with ASBJ Guidelines	*t*-Test (*P*-value) MWU-test Z (*P*-value)
Mean (std. dev.)	51.41 (10.358)	51.71 (11.126)	50.82 (8.911)	t = 0.284 (0.778)
N	51	34	17	z = 0.200 (0.841)
Missing data	4	2	2	
Total	55	36	19	
Maximum	84	84	65	
Minimum	38	38	38	

	Total n (%)	CPTA who agree with ASBJ Guidelines n (%)	CPTA who disagree with ASBJ Guidelines n (%)	Chi-square test (Sig)
GENDER				
Male	52 (94.5)	35 (97.2)	17 (94.4)	0.260 (0.610)
Female	2 (3.6)	1 (2.8)	1 (5.6)	
Total	54 (98.2)	36 (100.0)	18 (100.0)	
Missing data	1 (1.8)			
CERTIFICATE				
With certificate	54 (98.2)	35 (100.0)	19 (100.0)	–
Without certificate	0 (0.0)	0 (0.0)	0 (0.0)	
Total	54 (100.0)	35 (100.0)	19 (100.0)	
Missing data	1 (1.8)			
WORK EXPERIENCE				
Less than 1 year	0 (0.0)	0 (0.0)	0 (0.0)	1.893 (0.755)
More than or equal to 1–3 years	1 (1.8)	1 (2.8)	0 (0.0)	
More than or equal to 3–5 years	5 (9.1)	4 (11.1)	1 (5.3)	
More than or equal to 5–10 year	8 (14.5)	4 (11.1)	4 (21.1)	
More than or equal to 10–30 years	33 (60.0)	22 (61.1)	11 (57.9)	
More than 30 years	8 (14.5)	5 (13.9)	3 (15.8)	
Total	55 (100.0)	36 (100.0)	19 (100.0)	
POSITION				
Employee	9 (16.4)	7 (19.4)	2 (10.5)	0.723 (0.395)
Employer	46 (83.6)	29 (80.6)	17 (89.5)	
Total	55 (100.0)	36 (100.0)	19 (100.0)	
CLIENT				
Less than 20 SMEs	4 (7.4)	3 (8.3)	1 (5.6)	4.114 (0.391)
More than or equal to 20 and less than 50 SMEs	13 (24.1)	10 (27.8)	3 (16.7)	
More than or equal to 50 and less than 100 SMEs	14 (25.9)	10 (27.8)	4 (22.2)	
More than or equal to 100 and less than 300 SMEs	19 (35.2)	12 (33.3)	7 (38.9)	
More than 300 SMEs	4 (7.4)	1 (2.8)	3 (16.7)	
Total	54 (100.0)	36 (100.0)	18 (100.0)	

Table 6.1. *Descriptive statistics*

Panel A: AWARE

n (%)	Don't know	Know the name, but not contents	Know both the name and the contents	Total	Missing data
ASBJ Guidelines	0	2 (3.6)	53 (96.4)	55 (100.0)	0
General Standards	1 (1.8)	2 (3.6)	52 (94.5)	55 (100.0)	0

Pane B: IMPLEMENT

n (%)	Implemented	Not implemented	Total	Missing data
ASBJ Guidelines	22 (40.7)	32 (59.3)	54 (100.0)	1
General Standards	36 (81.8)	8 (18.2)	44 (100.0)	11

Table 6.2. *Awareness and implementation of SMEs accounting standards*

6.3. Results

First, our research investigated whether or not participants would agree with applying ASBJ Guidelines that incorporate the IFRS-based global accounting standards for SMEs. The reasons for their responses were also identified and codified. These investigations were performed to address RQ1 and RQ2 (see section 6.1).

Table 6.3 shows that 65.5% of participants (36 out of 55) in this study agreed with the idea that ASBJ Guidelines are necessary in addition to General Standards, while the remaining 34.5% contended that General Standards are enough and ASBJ Guidelines are not necessary to deal with the accounting for SMEs in Japan. Among the participants who agreed with ASBJ Guidelines, the largest response (25%) believed that ASBJ Guidelines are especially important for the SMEs that intend to list their shares in the financial market in the near future (IPO Purpose). Following this, the second largest response (13.9%) was that ASBJ Guidelines are applicable for relatively larger SMEs, whereas General Standards are applicable for smaller or micro SMEs. These respondents suggested that two different types of accounting standards for SMEs would be useful if the applicability of these standards could be segregated appropriately in accordance with firm size. Five respondents, however, did not clarify the definition of size as the criteria, but four other participants specifically indicated that size stood for the capacity size of financial departments (Resources) in the SMEs (two responses), sales (one response) and net assets (one response).

The third largest response of the participants (11.1%) reported that ASBJ Guidelines were regarded as the inevitable standard for the SMEs because of their benefits for financing activities (Financing Benefits). By applying ASBJ Guidelines, SMEs in Japan can enjoy the privilege of providing better financing conditions such as lower interest rates and fewer credit guarantee fees [KUS 14]. Other lesser responses agree with ASBJ Guidelines and indicate that the distinction between ASBJ Guidelines and General Standards is suitable to classify the structural types between public and family companies (Ownership structure), Accounting Advisor's Intention and Client's Intentions. These three reasons obtained two participants' responses each.

	Frequency (%)	Frequency (%)	
Participants who agree with ASBJ Guidelines	36 (65.5)		
		9 (25.0)	IPO purpose (Transition)
		5 (19.4)	SMEs' size (Not Specified criteria)
		4 (11.1)	Financing benefits
		2 (5.56)	SMEs' size (in Resources)
		2 (5.56)	Ownership structure
		2 (5.56)	Inquiry from accounting advisors (Advisor's Intention)
		2 (5.56)	Inquiry from clients (Client's Intention)
		1 (2.78)	SMEs' size (in Sales Amount)
		1 (2.78)	SMEs' size (in Net Asset)
		8 (22.2)	No specific reason
		36 (100.0)	
Participants who disagree with ASBJ Guidelines	19 (34.5)		
		5 (26.3)	Confusing
		2 (10.5)	Substantially same standards (Homogeneity)
		1 (5.26)	Guideline is too advanced
		1 (5.26)	Nuisance
		10 (52.6)	No specific reasons
		19 (100.0)	
Total	55 (100.0)		

Table 6.3. *CPTAs' perceptions of ASBJ Guidelines*

In contrast, among 19 participants who disagreed with the necessity of ASBJ Guidelines adaptation, the largest response (26.3%) for this question indicated that having two standards for SMEs would confuse both users and providers of financial statements. Followed by this response, two participants (10.5%) viewed the two standards as being substantially the same (homogeneity). Other minor responses suggested that ASBJ Guidelines should not have been applied because they are too advanced and a nuisance for SMEs.

Table 6.4 displays the CPTAs' perceptions toward values of applying ASBJ Guidelines as the SME accounting standards in Japan. Ten beneficial items investigated in the literature were incorporated in the questionnaire of the present study to address RQ3. Ten questions asked subjects to respond by using the five-point Likert scale method. This study also rated rank orders of mean scores responded from participants. Table 6.4 presents the information of ranking orders in each category of samples.

	Total (27)		Participants who agree with ASBJ Guidelines (19)		Participants who disagree with ASBJ Guidelines (8)	
	Mean (std. dev.)	Rank order	Mean (std. dev.)	Rank order	Mean (std. dev.)	Rank order
COMPARABILITY	3.19 (1.178)	5	3.05 (1.177)	6	3.50 (1.195)	1
RELIABILITY	3.70 (1.203)	1	3.79 (1.228)	1	3.50 (1.195)	1
TRANSITION	3.48 (1.451)	2	3.68 (1.416)	2	3.00 (1.512)	4
OVERSEAS FINANCE	2.30 (1.171)	9	2.32 (1.204)	9	2.25 (1.165)	8
DOMESTIC FINANCE	3.48 (1.252)	2	3.68 (1.293)	2	3.00 (1.069)	4
DIRECT INVESTMENT	2.30 (1.137)	9	2.32 (1.157)	9	2.25 (1.165)	8
EFFICIENT MANAGEMENT	2.96 (1.192)	7	3.05 (1.177)	6	2.75 (1.282)	6
EFFECTIVE M&A	3.00 (1.330)	6	3.11 (1.243)	5	2.75 (1.282)	6
EFFECTIVE AUDITING	3.37 (1.182)	4	3.47 (1.073)	4	3.13 (1.458)	3
EFFECTIVE TAXATION	2.37 (1.194)	8	2.47 (1.073)	8	2.13 (1.458)	10

Table 6.4. *Perceived values of adapting ASBJ Guidelines for SMEs*

The result reported that the highest mean score for all participants was RELIABILITY (3.70 for total; 3.79 for ASBJ Agreed Group; 3.5 for ASBJ Disagreed Group). As for the second and third top mean scores, TRANSITION (3.48 for total; 3.68 for ASBJ Agreed Group) and DOMESTIC FINANCE (3.48 for total; 3.68 for ASBJ Agreed Group) were similarly rated for both total and ASBJ Agreed Group, while COMPARABILITY was rated as another highest item for ASBJ Disagree Group (3.50) followed by EFFECTIVE AUDITING as the third highest item for this group (3.13). For ASBJ Disagree Group, TRANSITION (3.00) and DOMESTIC FINANCE (3.00) were both rated as the fourth highest items. On the contrary, the bottom three of the mean score were DIRECT INVESTMENT (2.30 for total; 2.32 for ASBJ Agreed Group; 2.25 for ASBJ Disagreed Group), OVERSEAS FINANCE (2.30 for total; 2.32 for ASBJ Agreed Group; 2.25 for ASBJ Disagreed Group) and EFFECTIVE TAXATION (2.37 for total; 2.47 for ASBJ Agreed Group; 2.13 for ASBJ Disagreed Group). Only participant group who did not agreed with ASBJ Guidelines rated COMPARABILITY a little higher (3.50) than the counterpart group (3.05).

	Total (37)		Participants who agree with ASBJ Guidelines (28)		Participants who disagree with ASBJ Guidelines (9)	
	Mean (std. dev.)	Rank order	Mean (std. dev.)	Rank order	Mean (std. dev.)	Rank order
COMPARABILITY	3.27 (1.194)	3	3.36 (1.224)	4	3.00 (1.118)	2
RELIABILITY	3.89 (1.265)	1	4.11 (1.227)	1	3.22 (1.202)	1
TRANSITION	2.65 (1.296)	8	2.71 (1.329)	8	2.44 (1.236)	7
OVERSEAS FINANCE	2.11 (1.100)	10	2.04 (1.105)	10	2.33 (1.118)	10
DOMESTIC FINANCE	3.81 (1.198)	2	4.11 (1.031)	1	2.89 (1.269)	3
DIRECT INVESTMENT	2.19 (1.076)	9	2.11 (1.100)	9	2.44 (1.014)	7
EFFICIENT MANAGEMENT	3.14 (1.294)	5	3.36 (1.339)	4	2.44 (0.882)	7
EFFECTIVE M&A	2.84 (1.214)	7	2.93 (1.245)	7	2.56 (1.130)	5
EFFECTIVE AUDITING	3.22 (1.357)	4	3.43 (1.372)	3	2.56 (1.130)	5
EFFECTIVE TAXATION	3.06 (1.453)	6	3.11 (1.577)	6	2.89 (1.054)	3

Table 6.5. *Perceived values of adapting General Standards for SMEs*

Table 6.5 shows the mean scores of perceived values to apply for General Standards responded by participants of this study. RQ4 was addressed by this analysis. This study also rated rank orders of mean scores as responded by participants. Table 6.5 presents the information of ranking orders in each category of samples.

	Paired samples *t*-test			Wilcoxon signed-rank test (WSR test)		
	ASBJ Guidelines mean (std. dev.)	General Standards mean (std. dev.)	*t*-Test (*P*-value)	ASBJ Guidelines mean rank	General Standards mean rank	*z*-value (*P*-value)
Frequency	23	23		22	22	
COMPARABILITY	3.22 (1.166)	3.35 (1.027)	–5.69 (0.575)	3.50	4.38	–0.604 (0.546)
RELIABILITY	3.70 (1.295)	3.83 (1.154)	–0.472 (0.641)	3.70	6.63	–0.483 (0.629)
TRANSITION	3.35 (1.526)	2.39 (1.196)	2.957 (0.007)***	6.67	3.00	–2.431 (0.015)**
OVERSEAS FINANCE	2.35 (1.191)	2.09 (0.996)	1.239 (0.228)	6.50	4.00	–1.218 (0.223)
DOMESTIC FINANCE	3.35 (1.301)	3.78 (0.998)	–1.447 (0.162)	3.38	6.92	–1.441 (0.150)
DIRECT INVESTMENT	2.26 (1.176)	2.17 (0.984)	0.463 (0.648)	5.30	4.63	–0.491 (0.623)
EFFICIENT MANAGEMENT	2.87 (1.217)	2.91 (1.259)	–0.204 (0.840)	4.30	5.88	–0.122 (0.903)
EFFECTIVE M&A	2.91 (1.411)	2.87 (1.217)	0.188 (0.852)	5.00	6.25	–0.262 (0.794)
EFFECTIVE AUDITING	3.30 (1.259)	3.00 (1.279)	1.127 (0.272)	5.57	5.33	–1.201 (0.230)
EFFECTIVE TAXATION	2.27 (1.077)	3.00 (1.309)	–2.203 (0.039)**	7.33	7.55	–1.946 (0.052)*

***Significant at the 0.01 level; **significant at the 0.05 level; *significant at the 0.1 level.

Table 6.6. *Differences in values between ASBJ Guidelines and General Standards*

According to this analysis, the top three of the highest mean scores in Table 6.5 were RELIABILITY (3.89; 3.22), DOMESTIC FINANCE (3.81; 2.89) and COMPARABILITY (3.27; 3.00) for both total and CPTA group that disagreed with ASBJ Guideline, respectively. As for CPTA group that agreed with ASBJ Guideline, RELIABILITY (4.11) and DOMESTIC FINANCE (4.11) were simultaneously rated as the highest rank, while EFFECTIVE AUDITING was rated as the third rank (3.43). COMPARABILITY was evaluated as fourth rank for this group (3.36).

In contrast, the bottom three of the mean scores were OVERSEAS FINANCE (2.11 for total; 2.04 for ASBJ Agreed Group; 2.33 for ASBJ Disagreed Group), DIRECT INVESTMENT (2.19 for total; 2.11 for ASBJ Agreed Group; 2.44 for ASBJ Disagreed Group) and TRANSITION (2.65 for total; 2.71 for ASBJ Agreed Group; 2.44 for ASBJ Disagreed Group).

Table 6.6 shows the comparison results for perceived advantages of implementing two different SMEs accounting standards among Japanese CPTAs. A paired sample *t*-test and Wilcoxon signed-rank (WSR) test were performed to compare their perceived advantages from 10 different perspectives addressed in various previous studies. This analysis was implemented to explore RQ5. The significant differences were found in TRANSITION ($t = 2.957$, $P < 0.01$ for *t* test; $z = -2.431$, $P < 0.05$ for WSR test) and EFECTIVE TAXATION ($t = -2.203$, $P < 0.05$ for *t* test; $z = -1.946$, $P < 0.1$).

6.4. Interpretation and discussion

6.4.1. *CPTAs' perceptions of ASBJ Guidelines*

This study first investigated to what extent CPTAs in Japan agreed with the necessity of ASBJ Guidelines in order to implement voluntary adoption of global comparable accounting standards and also identified the reasons for their perceptions. The findings suggested that the majority of CPTAs in Japan (65.5% of total participants) perceived the need for ASBJ Guidelines together with local GAAP of General Standards. In addition, the majority of CPTAs suggested that ASBJ Guidelines should be applied for SMEs to prepare for implementing IPO in the near future. Similarly, voluntary adaptation to global standards has been seen in emerging economies where some ambitious SMEs apply IFRS for SMEs to make their transition to full IFRS easier [KIL 14, UYA 13, MUL 10]. Given these studies, SMEs' global accounting standards, including IFRS for SMEs, tend to be thought of as the useful standard for SMEs to achieve their intensions for future IPO or upgrading to apply full GAAP for public companies. In this context, the result is consistent with previous studies. In Japan, public companies whose shares are listed in the financial markets need to adopt either J-GAAP for Public Companies, JMIS (Japan's Modified International Standards: Accounting Standards Comprising IFRSs and the ASBJ Modifications) or full IFRS. Adopting these higher quality standards however requires them to have more advanced internal accounting systems and invest more of the firm's resources and time. Thus, it is interpreted that CPTAs in Japan perceive that SMEs' choice of adapting ASBJ Guidelines is an appropriate first step to narrow the gaps in accounting development between SMEs and public companies. With regard to this interpretation Uenishi [UEN 12a] also stressed in his research that CPTAs in Japan are responsible for encouraging SMEs which achieved quality of General Standards to shift toward more advanced ASBJ Guidelines. The finding of this research could be presented due to their responsible attitude.

This chapter also found that the second largest group of CPTAs contended ASBJ Guidelines is suitable for relatively large SMEs to deal with their accounting. As seen in the literature review section, previous studies empirically examined the effect of firm's size as the predictor for voluntary choice of global accounting standards. But they concluded that these outcomes vary in each study and did not reach a consensus [EIE 13, ALB 13b, AND 12, LIT 12, QUA 12, CAR 10]. Furthermore, it was found that 13.9% of participants failed to define what the firm's size means, although they agreed to use size to distinguish SMEs' application between global ASBJ Guidelines and local General Standards. This could be a major weakness of our study which did not employ statistical analysis of correlation between firm size and other proxy variables with using Japanese CPTA settings. Prior studies investigated firm size because this criterion was thought to as the proxy for other determinant variables for voluntary disclosure of global comparable financial information. These variables were total assets [EIE 13], numbers of employees [EIE 13], information production cost [AND 12], financial resource [AND 12] and international activities [AND 12]. Some minor responses of the current study defined firms size as resources of finance department, sales and net asset, but it is suggested that other major participants who responded firm size should have expressed more accurately what the size of the SMEs would stand for.

Out of 36 participants who agreed with ASBJ Guidelines, four (11.1%) suggested that ASBJ Guidelines were an appropriate standard to obtain better privileges in finance from financial institutions. This aspect can be an advantage for applying ASBJ Guidelines, so that it may be consistent with the effect of reducing SMEs' capital cost [KUS 14, MUR 10]. However, this advantage was limited in a domestic sense, and overseas finance such as reaching global capital was out of bounds.

Ownership structure was simply a minor response from CPTAs in Japan. Also, no response regarding international activities was found in this study. These findings were highly contrasted with active discussions of prior studies in overseas [EIE 12, LIT 12, BAS 12], where significant correlations were found between voluntary adaptation of global accounting standards and active international finance and trades.

6.4.2. *CPTAs' perceived values of ASBJ Guidelines*

Second, this chapter explored the CPTAs' perceptions regarding values of applying ASBJ Guidelines as SME accounting standards in Japan. According to the analysis outcomes, the findings revealed that CPTAs in Japan strongly believed that ASBJ Guidelines work effectively to enhance the reliability of SMEs' financial reporting (RELIABILITY), help SMEs transition smoothly to full accounting standards for public companies (TRANSITION) and enable them to have better finance in the domestic circumstance (DOMESTIC FINANCE). The latter two results are especially consistent with the findings in Table 6.3, where ASBJ Guidelines were appreciated due to their positive impact on implementing a smooth transition to full GAAP and for providing better finance treatments in Japan.

On the other hand, lower mean scores of DIRECT INVESTMENT and OVERSEAS FINANCE confirmed the CPTAs' perceptions that ASBJ Guidelines do not have any advantages for overseas activities such as overseas finance and direct investments, although this standard has been incorporated within the IFRS framework. Several studies in Japan highlighted that ASBJ Guidelines had been subjected to IFRS influence and regarded to as the global comparable reporting standards (e.g. [KAW 12]). However, the finding of this study demonstrated that actual practitioners of tax accounting in Japan who normally engage with SMEs' management practices experienced the negative effects of ASBJ Guidelines upon their activities overseas. This is a unique feature of SMEs' accounting scheme in Japan. Overseas studies reported that international reporting strategy using International Accounting Standards or US GAAP directly improved economic value measured by accounting figures (e.g. [LEU 00]). However, the study showed inconsistent results that CPTAs who get involved in

SMEs' accounting perceived internationalization strategy, in terms of accounting, not to be economically valued.

Although the rank order showed little difference in COMPARABILITY, these differences between CPTAs who agreed with the necessity of ASBJ Guidelines and those who disagreed did not have a big impact on interpretations about perceived values of ASBJ Guidelines. It is concluded that two groups of CPTAs in Japan shared similar perceptions toward the values of ASBJ Guidelines regardless of their agreement on this standard.

6.4.3. *CPTAs' perceived values of General Standards*

Third, this chapter also evaluated perceived values of General Standards among CPTAs. The score of TRANSITION in Table 6.5, that was rated in the top three items, was very contrasted in comparison to the scores in Table 6.4 where the same item of TRANSITION was rated as one of the bottom three items. According to this outcome, it is interpreted that CPTAs perceive that the effective transition to the full GAAP would not be succeeded by General Standards. This result is consistent with the previous studies articulating that ASBJ Guidelines are more complex and closer to the full GAAP for public companies than General Standards (e.g. [KAW 14, KAW 12]). Rather than the transition function, the finding demonstrated that participants of the study highly evaluated more qualitative function, which implies that General Standards work effectively to add better reliability and comparability to financial statements. In this regard, Kawasaki [KAW 12] also indicated that General Standards are regarded to reflect specific features of SMEs in the standard setting process and avoids influences from large entities' GAAP. Accordingly, our study ensured that CPTAs in Japan consent with prior studies and view that General Standards would be well suited for SMEs financial reporting purpose.

Two variables, OVERSEAS FINANCE and DIRECT INVESTMENT, were rated the lowest rank orders by both participant groups. This finding can be interpreted that both CPTA groups agreed that the General Standards would be less effective for the SMEs in Japan to enhance global financial and business activities, regardless of their acceptance to the ASBJ Guidelines. This perception among CPTAs in Japan is consistent with the specific feature of General Standards that allows SMEs to avoid completely the impact of IFRS [KAW 12].

6.4.4. *Differences in value between ASBJ Guidelines and General Standards*

Finally, this chapter compared the differences in CPTAs' perceptions between values of ASBJ Guidelines and those of General Standards. This enabled us to evaluate CPTAs' views regarding particular values of each standard. The findings of the analysis indicated that participated tax accounting professions in Japan evaluated ASBJ Guidelines as the SMEs accounting standards more strongly than General Standards in terms of preparing for the transition to the full GAAP scheme. On the other hand, our participants evaluated General Standards as more effective standards to deal with tax accounting practices than ASBJ Guidelines. It found no significant differences in the scores of other eight variables between the two SME accounting standards.

The role of ASBJ Guidelines in the transition process was also evident in other parts of the outcomes in this research (findings in Tables 6.3 and 6.4). This statistical finding again supported the strong and unique features of ASBJ Guidelines (e.g. [UEN 12a, UEN 12b]). In addition, this benefit to enhance SMEs' future transition is interpreted not as the economic value that is directly associated with the short-term improvement of stock performance and lower cost of capital, but as a sustainable long-term value that related with a wider range of stakeholders and company environments.

The role of General Standards for effective taxation has been also articulated in the literature. Murata [MUR 10] indicated the primary reason for SMEs to avoid applying ASBJ Guidelines is due to a strong bond between financial and tax accounting referred as the principle of congruence. Uenishi [UEN 12a, UEN 12b] also stated that General Standards allow SMEs to apply the accounting standards that are basically prescribed by taxation law. The author analyzed that this harmonized trait with taxation scheme attracts SMEs to apply General Standards more than ASBJ Guidelines. The outcome that the role of EFFECTIVE TAXATION for General Standards evaluated significantly by participated CPTAs more strongly than that of ASBJ Guidelines is consistent with this context. Compared with the long-term sustainable value from the transition purpose of ASBJ Guidelines, this benefit from effective taxation for General Standards is somehow regarded as the economic value that contributes directly to reducing the SMEs' cost of taxation.

6.5. Conclusion

The objective of this chapter was to investigate the determinant factors affecting to Japanese SMEs' accounting choice regarding voluntary disclosure of global comparable information especially in order to understand how their decisions link with the firm's value. To achieve this research aim, the perceptions of SME accounting standards among CPTAs in Japan were examined.

The findings concluded that the majority of CPTAs in Japan believed that IFRS-based global ASBJ Guidelines should coexist with local GAAP-based General Standards. The reasons for their perceptions were also addressed and it was found that CPTAs tended to view ASBJ Guidelines as a beneficial standard, particularly for SMEs whose size was relatively larger, where managers intended IPO or smooth transition to full GAAP, and desired better privileges in finance. Furthermore, our investigation of perceived values for two SME accounting standards revealed that CPTAs believed ASBJ Guidelines do not have any advantages for overseas activities including cross-border financing and trade, of which finding was highly contrasted with prior studies in other countries.

This chapter contributes to the literature by illustrating one case study of voluntary disclosure in a developed country like Japan. The findings of this study substantially answered to the calls by previous studies for further research regarding SME accounting standards in developed countries [KAY 15, UYA 13]. It was found that SMEs would be motivated to apply global accounting standards not because they intended to expand their cross-border activities, but because they rather tended to concentrate on enhancing more domestic finance and trade. This unique trait reflected CPTAs' perceptions supporting IPO and finance purpose as the values of ASBJ Guidelines, while its roles of overseas finance and direct investment were evaluated lower. This could be attributed to the fact that capital cost for SMEs in Japan to access to the domestic finance will not be higher compared with finance situations for SMEs in other countries. Without demands of cross-border finance in SMEs in Japan, their purpose of preparing SMEs financial reports would focus more on improving quality of reporting in the domestic financial context rather than providing global comparable information. The findings of this research will be of great interests for SME accounting standard decision-makers such as the International Accounting Standards Board (IASB), ASBJ and SMEA for preparing appropriate accounting regulations and standards for SMEs in the future.

Although our findings provided several contributions to the literature, these are not free from limitations. Among them, one of the major limitations was a lack of definition for firm size as the effective criterion for applying ASBJ Guidelines. Effectiveness of firm's size as the definition of SMEs has been discussed in previous studies (e.g. [EIE 13, AND 12, LIT 12, CAR 10]), but the research findings are contradictory and inconclusive. In this chapter, firm size was viewed by CPTAs as one of the influential drivers to determine application of ASBJ Guidelines, but participants did not clearly identify what firm size stands for. The questionnaire-based survey applied in this chapter only allows us to obtain limited amounts of information. Qualitative research methodology such as in-depth interview and ethnography could be conducted in further research to address the weakness of this study. Additionally, this issue can be similarly addressed from other stakeholders' point of views including SME managers, policy makers and financial providers. Further research efforts in this field would help improve the quality of SME accounting in the future.

7

SMEs' Value from Voluntary Disclosure: Inductive Qualitative Approach

7.1. Introduction

This chapter performs inductive research with the exploratory analysis that determines the factors predicting voluntary disclosure of financial statements by SMEs. Inductive research generally begins with observations and seeks to identify relations and patterns from which theories can be developed [HYD 20]. The analysis technique applied for this research was thematic analysis, which was defined by Braun and Clarke [BRA 06] as the method for identifying, analyzing and reporting patterns as the themes within data. More specifically, this research carried out a theoretical thematic analysis [BRA 06, BOY 98]. This type of thematic analysis would tend to be driven by the researchers' theoretical or analytic interest in a specific area such as the research question in this current study. The deductive type of theoretical thematic analysis was thought to be less suitable for providing a rich description of the data overall but rather for analyzing detailed aspects of the data [BRA 06]. Given this contextual background, this research employed the thematic analysis method illustrated by King and Horrocks [KIN 10].

In this chapter, a qualitative in-depth semistructured interview approach was applied with broad, exploratory and open-ended questions, including prompts to obtain more depth in the response and to work collaboratively with the respective participants [KIN 10]. For the interviews, an interview schedule was used with the primary question seeking the detailed process used by SMEs when choosing accounting standards, together with demographics of interviewees. Further, the degree to which Certified Public Tax Accountants (CPTAs) and SMEs collaborate in order to deal with firms' accounting matters was also addressed by asking both CPTAs and SME senior managers their viewpoints, respectively. A little adjustment was made to the interview schedule based on the first analyses of the data from CPTAs to address important areas of interest in more depth. For this purpose, senior managers were additionally asked questions on future perspectives or vision for their companies to clarify whether or not these aspects would affect their choice of accounting standards.

Face-to-face in-depth interviews lasted 20–50 min for CPTAs and 40–60 min for senior managers. Apart from two interviews of S1 and S4, all other interviews were digitally audio-recorded and transcribed verbatim in the Japanese language. For the two interviews without audiotaped data (S1 and S4 in Table 7.1), contextual details were recorded in journals immediately after each interview.

7.2. Research design

7.2.1. *Participants*

Participants of this research consisted of Japanese SME senior managers such as CEOs, CFOs and employees who were primary responsible for accounting and financial issues within their companies. Additionally, Japanese CPTAs were included in our research sample. CPTAs were regarded as eligible because the recent survey revealed that 65.8% of SME senior managers did not exactly know the accounting standards they applied. Instead they relied on advice from external accounting professions including CPTAs who were hired by SMEs as tax and accounting advisors [TEI 13]. These participants were selected on purpose so as to capture the perspectives of SME top managers from various industries, firm organizational structures, firm sizes as well as whether they are obtaining loans or not. CPTAs also came from a variety of positions with differentiated work experience, number of clients, age, gender and size of accounting firms. In total, 12 interviews were conducted among six CPTAs' and 10 SMEs' top managers and accounting staff (see Table 7.1). Of the 12 interviews, five CPTAs (TA1, TA2, TA3, TA4 and TA5) and three CEOs (S4, S6, S7) were interviewed individually, while another four interviews (S1, S2, S3 and S4) were carried out as a group including CEOs, CFOs, other staff and CPTA employed by SMEs.

Panel A: CPTAs

Interview ID	TA1	TA2	TA3	TA4	TA5
Interview date	July 13, 2015	September 1, 2015	September 1, 2015	September 1, 2015	August 4, 2015
Interview time (min)	28	20	43	29	48
Position in accounting firm	Employer (A)	Employer (B)	Employer (C)	Employer (D)	Employer (E)
Gender	Male	Female	Male	Female	Male
Age	40s	30s	40s	40s	60s
No. of clients (firms)	50–100	500–1,000	100–150	50–100	50–200
Work experience (year)	17	10	15	19	27
Size of firm (employees)	1–5	20–50	1–5	5–10	5–10
Transcribed	Yes	Yes	Yes	Yes	Yes
Total numbers of codes	56	38	51	58	47

Panel B: SMEs' Top Managers

Interview ID	S1	S2	S3	S4
Interviewee	CEO (F) Director (G)	CEO (H) CPTA (I)**	CFO (J) Director (K)	CFO (L)
Interview date	September 3, 2016	February 18, 2016	August 8, 2016	August 9, 2016
Interview time (min)	60	47	60	40
Industry	Manufacturing (MF)	Chemical	MF	MF
Structure	Owned by managers (OBM)	OBM	Totally hold subsidiary	OBM
SME size	Large	Middle	Large	Middle
Capital (¥1,000)	320,000	20,000	480,000	30,000
Annual turnover (¥1,000,000 in 2015)	33,000	332,000	17,600	1,900
Employees (in 2015)	847	8	948	78
Debts (¥1,000)	60,000	NA	0	Minus
Accounting standard	Tax Accounting (TA)	J-GAAP	J-GAAP	ASBJ Guidelines
Transcribed	No	Yes	Yes	No
Total numbers of codes	10*	53	41	10*

Interview ID	S5	S6	S7
Interviewee	CEO (M) Director (N)	CEO (O)	Director (P)
Interview date	August 8, 2016	August 9, 2016	August 9, 2016
Interview time (min)	55	50	60
Industry	Real estate	Advertising	Brewery
Structure	OBM	OBM	OBM
SME size	Middle	Micro	Small
Capital (¥1,000)	12,000	15,000	60,000
Annual turnover (¥1,000,000 in 2015)	600	270	350
Employees (in 2015)	85	17	23
Debts (¥1,000)	0	130,000	Minus
Accounting standard	General Standards (GS)	GS	TA
Transcribed	Yes	Yes	Yes
Total numbers of codes	45	10*	43

*Data in the journal were codified because tape recording was not permitted.

**Participant I was also registered as Certified Public Accountant and a member of Japanese Institute of Certified Public Accountants (JICPA).

Table 7.1. *Participants' demographics*

This research collected data from 12 interviews in line with the findings of [GUE 06], who indicated that 92% of the total number of codes could be created and new themes would not emerge frequently and/or progressively after an analysis of 12 interviews. This finding is eventually subjected to the consensus theory advocated by Romney *et al.* [ROM 89] that justified the principle that experts tend to agree more with each other (with respect to their particular domain of expertise) than do novices particularly under three assumptions: existence of external truth in the research domain, independency of participants in the interview and coherence of domains of knowledge in questions. The current research setting supported these assumptions since our participants shared common experiences of standard choice in actual business practices (external truth), conducted their interviews separately (independency) and responded to similar questions with relatively narrow objectives (coherent knowledge domain).

Accordingly, 12 interviews were regarded as a sufficient number to conduct the qualitative analysis referred in section 7.2.2.

7.2.2. *Data analysis*

In the analysis process, this chapter reads each transcription closely, then creates preliminary comments line-by-line while defining the label for each comment. This is referred to as a descriptive coding step [KIN 10]. In this study, the descriptive code is called "contextual comments". This was followed by an interpretative coding procedure, where the descriptive codes were clustered and the meaning of these clusters interpreted in conjunction with the research question [KIN 10]. This interpretative code is also called "components". The final step of the analysis process was to define a number of overarching themes. Such themes were identified by constructing interpretative codes and were reviewed and refined in this step. The final form of the themes should be at a higher level of abstraction than the interpretative codes [KIN 10]. In this chapter, we construct several subthemes associated with the main theme. A computer software package known as Atlas ti. was used to deal with this whole coding process.

7.2.3. *Quality check*

In qualitative research, including the thematic analysis method, it is argued that there is no general agreement about which criteria to use when assessing quality or how to apply the qualitative methods to the criteria that are normally used for quantitative research [KIN 10]. Along with this situation, this research attempted to employ one of the influential criteria advocated by Guba and Lincoln [GUB 89]. Their set of criteria contains four quality criteria as the direct alternative to the main criteria used in quantitative research: credibility, transferability, trackable variance and confirmability. The data in this study were reflected in the terms of all four of these criteria.

7.3. Results

The most important themes emerging from the analysis are summarized below. Three themes were identified from the data and are shown in Figure 7.1. These themes were (1) accounting competencies, (2) financing needs, and (3) SMEs' management goals. They were illustrated by quotes that were translated from Japanese into English, which were carefully edited to make them more readable without any loss of meaning.

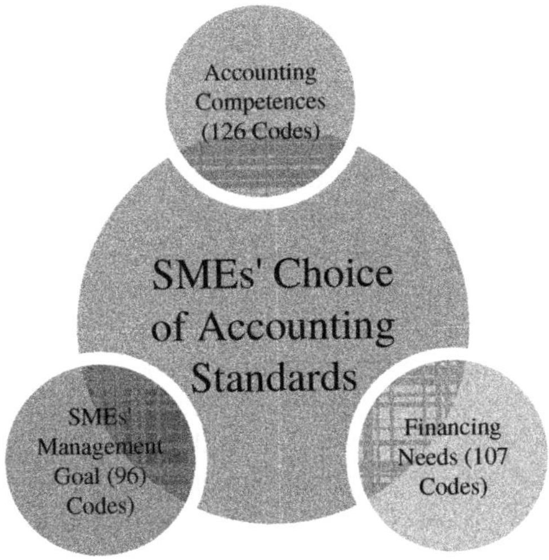

Figure 7.1. *Themes affecting SMEs' choice of accounting standards for voluntary disclosure*

All three themes emerged as the seminal factors affecting SMEs' choice of accounting standards for voluntary disclosure. The first theme was described as "accounting competencies". This theme highlights the quality of accounting competencies developed in the SMEs that affect the processes of constructing in-house accounting systems: personnel in the financial department and relationships with the external accounting profession. The second theme was labeled as "financing needs", which refers to the SMEs' financial situation in terms of whether or not they need bank finance. The third theme was defined as SMEs management goals and this relates to the SMEs' management attitude, to the firm's positive outlook as well as to the various stakeholders who may demand financial statements.

7.3.1. *Accounting competencies*

This theme can be split into two subthemes that are characterized by interaction with external accountants – the type of SME that is totally independent from the external accounting profession and the one that is highly dependent on the accounting profession.

7.3.1.1. *Accounting practices independent from external accounting profession*

Some participants of this research admitted that the firm that is free from influence of the external accounting profession has substantial autonomy to make a choice on which accounting policy to adopt. This type of flexibility in accounting is due to highly qualified personnel in the accounting departments, which is typically observed in the large SMEs such as S1 and S3. In these firms, many competent employees are found in the accounting division. Accordingly, their in-house accounting personnel are capable of preparing the firm's own financial statements independently and so avoid the necessity for advice from the external accounting profession. K in S3 stated this aspect as follows:

> "In my company, the finance department is responsible for all accounting transactions and related issues including lodgment of taxation returns. An external CPTA is hired by us just in case unexpected situations arise. Normally we simply ask the CPTA to double check our work. So basically we do all our accounting by ourselves".

Similar comments appeared in other participants' accounts (C in TA3, D in TA4 and E in TA5). Accordingly, SMEs' autonomy of accounting standard choice will be facilitated in line with the fully fledged in-house accounting department that comprises competent staff.

In contrast, other SMEs which do not have sufficient human resources can also enjoy this autonomy of accounting choice by adopting an accounting system that is supported by information technology. Such companies, which are often smaller sized SMEs, have a low budget to allocate to the accounting section and thus do not tend to hire external professional accountants. The participants in S5, for instance, have not appointed any full-time employees in their accounting section. Instead, there were three part-time accounting staff and they worked with the support of information technology. Using this system, all accounting transactions were automatically gathered, and these part-time employees regularly input aggregate data into the accounting software to prepare financial statements. The interviewer M (CEO) of S5 explained as follows:

> "My company has tended to develop a hands-free accounting system that basically allows any employees to manage our accounting. We are reluctant to spend additional funds to hire

> full-time accounting staff. Also, we don't purchase any expertise from the external accounting profession. Once we headhunted an employee who came to us from a local tax accounting firm where he worked in corporate taxation. Our accounting goal is to reduce our taxation liability. My company does not clearly identify which accounting standards are to apply, but I guess in a way we are using the General Standards".

In this firm, although being smaller sized and less resourced, the high-quality in-house financial department has been developed by installing information technologies that allow the firm not to involve the external accounting profession in the firm's accounting practices. As a result, this firm is still maintaining autonomy in accounting.

7.3.1.2. *Accounting practice dependent on external accounting professionals*

It was found that SMEs highly dependent on external accountants would behave differently in their choice of standards compared to SMEs who are less influenced by external professions. From our interviews, this difference was first presented to us by the indication that employees had little confidence in their accounting knowledge and competences. Participants in this category felt that senior managers were relatively pessimistic about their lack of accounting knowledge and skills. Participant H in S2 illustrates this proposition as follows:

> "Managers in SMEs are always preoccupied by many management issues including finance, sales, employment and so on. Complex accounting issues surrounding taxation and other related areas are beyond the ability of CEOs. In addition, we [as SME managers] do not have sufficient accounting knowledge because of these reasons, I think that most SME managers do not realise that there are opportunities to actually choose accounting standards".

Further, they admitted that they lacked the human resources to deal with accounting matters in the firm but nor could they afford to develop in-house staff competence in accounting. As a solution, SMEs in this category were more inclined to headhunt outsiders who had been working in the accounting and finance fields such as from accounting firms and commercial banks. Due to such deficiencies in their in-house expertise and resources, it has been observed that the SMEs in this category rely highly on the professionals to seek advice particularly when it comes to the choice of accounting standards. This component is illustrated in Professional Support (Table 7.2). This was highlighted by the following comment from participant H of S2:

> "Since I am working with my professional accountant (as a business advisor), I believe his work should be perfect, so I think it will be rude if I would add my comments or ask further questions. Or perhaps I am afraid if this professional, when I oppose him, replies to me 'what are you talking about since you don't have a good accounting knowledge?' This may be the untouchable part, and all other companies may fully rely on the accounting professionals for advice without question. In my case dealing with some part of the accounting myself would be rare. Other companies may surrender to the professional's opinions. This isn't my complaint but rather I try to express my opinions to the profession…I would say 'why don't we do this next year because of this or that…' I always have this kind of conversation with my accountant".

Technically speaking, the choice of accounting standards should be made by SMEs' top managers but it is reported from interview analysis that accounting professionals often have certain room for suggestions or exert direct influence with regard to the firm's choice of accounting standards in practice.

The last component in this subtheme was about professionals' attitudes toward accounting (Table 7.2). Two opposing constructs were found in this component. SMEs would choose either of two accountant types as their advisor. On the one hand, there are the accountant groups who believe that applying accrual accounting is the appropriate choice of Standard for SMEs. In this group, either ASBJ Guideline or General Standards can be chosen by the accountants. Other accountant groups could be more likely to think that accrual-based standards are rather harmful to SMEs. They believe financial statements should be a by-product after tax returns are completed. The SMEs that procure advice from accountants in this group are more inclined to choose Tax Accounting Standards rather than applying either of the two financial accounting standards.

Participant C of TA3 expresses the importance of accrual accounting as follows:

> "I am stressing the importance of developing in-house competencies to keep accounting records more timely and precise. Accounting can be used as the compass of the journey, and [SMEs] won't be guided properly without this compass. Recently, top managers have been required to explain their company's financial conditions to the banks using their own words [for finance purpose]. SME financial statements have been prepared in compliance with tax-based accounting standards. With the Corporate Tax Act, regular periodic depreciation is not necessarily adopted, but this accounting treatment fails [SMEs] to present a true view of the firm. We should seek drastic changes in a firm's accounting policy where it does not comply with the Corporate Tax Act. I guess quite a few number of CPTAs recognize this importance [in financial accounting for SMEs]".

The majority of CPTA participants in this research similarly expressed a positive attitude toward applying accounting standards reflecting accrual accounting principles, while one CPTA was skeptical of the effectiveness of accrual accounting. Participant E's account of TA5 illustrated this passive view regarding accrual accounting.

> "SMEs can use their financial statement only for banks and the tax office! For whom would they use adequate accounting methods and prepare financial reports? Nobody! [SMEs] are now applying [either ASBJ Guidelines or General Standards] because they can receive a reduction of the interest rate. But they would stop [applying these two accounting standards] if this scheme was to be abolished. […] The idea [of applying accounting standards for SMEs] is to urge SMEs' income statements to be more useful. […] But this does not work well, because there is no one that SMEs have to report [their financial statements] to. SMEs' financial reporting will be demanded by our society if it is necessary but it won't. [SMEs] do not have external shareholders. Rather their interests lie with the tax office [via their tax return] and finance availability from their banks. All CPTAs basically determine a client's bottom line in accordance with such interests [tax return and finance]".

Such dichotomous attitudes toward accounting by SMEs appear to be substantially affected by individual backgrounds and knowledge of SMEs accounting standards. It was found that some participants who positively evaluated SMEs accrual accounting belonged to a professional body, where members were frequently provided with opportunities to learn more about General Standards. Other groups however appeared to fail to recognize ASBJ Guidelines and General Standards as the independent set of accounting standards. As seen above in E's account, they are more likely to regard the choice of these standards simply as the tools to achieve the values available to them in the finance area. Consequently, this subtheme indicates that the choice of accounting standards in this category will be highly influenced by the accountant's professional attitudes. Who will be chosen as an accounting advisor by SMEs will be a key when formulating their accounting behavior especially in regard to the choice of standards.

Subthemes	Components	Contextual comments from interviews	Interview ID
Accounting practice independent from external accounting professions	High-quality in-house financial department (11 Codes)	Finance division of the firm treats all accounting matters.	S2
		The firm has been fully equipped with in-house competencies to deal with accounting issues including taxation issues.	S3
		The firm has invested and developed a high-quality IT system for accounting practices, instead of hiring professionals.	S5
	Limited responsibility of professions (15 Codes)	Responsibility of the professionals hired by firm is limited to taxation matters only.	TA1, S2, S3
		The firm does not hire accounting professionals.	S5
Accounting practice dependent on external accounting professions	Lack of accounting expertise and resources (32 Codes)	The SMEs managers are not good at dealing with numbers and accounting issues.	TA4, TA3,
		The SMEs managers do not know which accounting standard is applied within their firm.	S2, S6, TA3
		The SMEs managers are suffering from lack of accounting knowledge and skills	TA2, TA3, TA4, TA5, S6, S7
		Lack of eligible human resource who can manage accounting is a serious issue.	S2

		Staff in accounting division are headhunted by banks and other financial institutions.	S7, TA1, TA4
	Professional support (34 Codes)	Professionals assist SMEs with creating financial statements.	S2, S6, S7, TA2,
		Professionals do not make decisions on accounting matters, but support manager's decision.	TA1, TA2, TA3, TA4
		Professionals always chose adequate accounting standards applicable to the firm.	S2, TA4
	Professionals' attitude toward accounting (34 Codes)	To apply appropriate accounting principles is important.	TA1, TA2, TA3, TA4, TA5, S2
		To apply appropriate accounting principles is useless.	TA5, S2
		Tax accountants basically take on complex professional responsibilities.	TA4
		Professionals do not recognize specific accounting standards such as General Standards and ASBJ Guidelines.	TA2, TA3, TA5

Table 7.2. *Theme of accounting competences*

7.3.2. *Financing needs*

This theme will be divided into two subthemes defined by the SMEs' intention as to whether or not they require finance and whether this aspect is thought to have a strong impact on voluntary disclosure for SMEs.

7.3.2.1. *SMEs that need to finance*

Given the analysis of our interview data, it was found that the level of SMEs' demand for loans from banks drives their choice of accounting standards in the following two directions. First, the SMEs that are willing to borrow money from banks will have a strong incentive to comply with certain accounting standards because the healthy condition of financial outcomes and position are believed to help them obtain finance from the banks (S2, S4 and S6). In such a case, top managers of SMEs strive to prepare and report their firm's financial statements in a sound manner by means of applying accrual accounting rather than Tax Accounting Standards.

They have an incentive not to manipulate profit to decrease income tax payment, but to demonstrate a better shape of the company from a financial accounting perspective. This perception was observed in participant A's account of TA1. He explained as an advisor why SMEs tend to report an accurate view of the company by using accrual accounting when in need of loans from banks:

> "Thinking of SMEs' relationship with the banks, I highly recommend my clients to pay some tax and try to increase their retained earnings. The banks often won't accept a loan contract immediately after SMEs apply for it. I don't know what other CPTAs suggest but I suggest my clients pay their proper tax due each year. It is because the companies are supposed to last 5 to 10 years, so they need to manage a good relationship with their banks".

From this above account, it is also clear that advice from external accountants may be influential on an SMEs' choice of accounting policy. Participant A eventually indicated the possibilities that other CPTAs suggested differently. Although the majority of participants in this study agreed with the context of this comment, it is important to be aware that this consequence may be subject to external professionals' attitudes toward accounting as seen in section 7.3.1.2.

Second, it is interpreted that an SME's actual choice of accounting standards between ASBJ Guidelines and General Standards will be determined by the impact of the checklist schemes. The type of checklist scheme is twofold. As the first line, the Japan Federation of Certified Public Tax Accountants' Associations (JFCPTAA) published a checklist for ASBJ Guidelines to assure adequate application of the Guidelines for SMEs' financial reports and many commercial banks are now introducing special loan services using this checklist. With regard to this service, if SMEs successfully show evidence that they meet the criteria for the ASBJ Guideline by ticking all questions in the checklist then the loan collateral is exempted when SMEs borrow money from the bank. Furthermore, as the second line, JFAPTAA similarly issued a checklist for General Standards to assure application of its standard for SMEs' financial report. To promote this standard among SMEs, it was proposed by the Financial Service Agency (FSA) and the Small and Medium Enterprise Agency (SMEA) that the fee for the credit guarantee will be discounted 0.1% of the principled amount of the loan, when SMEs are assured to apply General Standards properly and tick all items to indicate compliance in the checklist. Given these checklist schemes, for those SMEs that especially need finance, their choice of standards between ASBJ Guidelines and General Standards will be highly subject to SMEs demands in regards to the financial values throughout these checklist schemes. More precisely, accounting professionals will be involved in this decision process as the advisors of the SMEs. Participant B of TA2 indicates as follow:

> "[…] It is interesting to know that the interest rate of the loan will be discounted if the company completes the checklist for General Standards. I know some managers know this fact, but others don't, so I try to encourage as many of my clients as possible to go with the General Standards so that they enjoy at least a little of the benefit when they obtain a loan. I know running a business requires lots of money".

As for the relationship between checklist schemes and choice of accounting standards, all CPTA participants expressed concerns about unintended consequences. They indicated that banks' inappropriate request for submitting the checklist caused distortion of the correct choice for accounting standards among SMEs. This issue is clearly illustrated by participant I's account interviewed in S2:

> "Basically I intend to help my client firms apply proper accounting because I am a Certified Public Accountant. Because of this reason, I normally apply J-GAAP with my clients. Then, bankers request me [and my clients] to submit this [checklist, when my clients require a loan]. So I thought that perhaps the accounting [of my clients] would be qualified, not equivalent but similar, as that of ASBJ Guidelines [in a sense that ASBJ Guidelines is more advanced and closer to J-GAAP than the General Standards]. But the bankers didn't know [about ASBJ Guidelines] and said that the client's loan can't be accepted unless I agreed to use the General Standards checklist. I was so disappointed because I know General Standards is the lowest quality of standards. Of course I modified the contents of the checklist [suitable for General Standards] because my client's accounting is obviously qualified for [a lower quality of] General Standards, too. […] But I found it is a very strange convention used by the banks".

Several participants in this study contended that the lack of in-depth understanding from the lenders' side on SMEs accounting standards causes a distortion in choosing the correct standards. They also suggested that this wrong choice of standards affected by the checklist resulted in a lower quality of information about the financial position and performance of the firm.

7.3.2.2. *SMEs that do not need finance*

In contrast to the case for SMEs that require finance, SMEs that do not need finance from capital providers have no pressure on their disclosure of financial information. These SMEs have autonomy as to whether or not they prepare financial statements because they are only mandatorily required to comply with the Corporate Tax Act for tax return lodgment. Reporting financial statements is not legally demanded of SMEs if they are not financed by the equity market.

Subthemes	Components	Contextual comments from interviews	Interview ID
SMEs that need finance	Accrual accounting (28 Codes)	There is the positive correlation between financing requirements and profit.	TA1, TA2, TA4
		SMEs should contribute to the society by paying tax as the result of earnings.	TA1
		Manipulating profit for taxation purpose is not beneficial to SMEs.	TA3
		When SMEs suffer from deficit, they are mostly concerned about bank requirements.	TA5
	Submitting checklist to the banks (59 Codes)	Financial institutions do not categorically understand the contents of accounting standards such as General Standards and ASBJ Guidelines.	TA1, TA2, TA5, S2
		Credit guarantee fees or collateral would be discounted if SMEs submit the checklist for accounting standards to the commercial banks.	TA1, TA2, TA4, TA5, S2, S4
		Pressures to amend checklist from banks or SMEs sometimes go further from the appropriate practice in accounting.	TA1, TA3, TA4, TA5, S2
		Choice of accounting standards is somehow driven by the banks.	TA3
SMEs that do not need finance	No pressure from bank to borrow the money (20 Codes)	No pressure is received from banks because the firm does not borrow money from them.	S1, S3, S5

Table 7.3. *Themes of financing needs*

Further, it was also found that the size of the firms does not correlate with their choice of standards. For example, one large SME interviewed in this research chose not to apply any accounting standards and only submits tax returns (S1), while the other smaller SME in our sample voluntarily applied General Standards for managerial purposes (S5). They were commonly owned by limited stakeholders (family and/or managers) and they also did not experience any pressure from the banks due to their financial situations.

7.3.3. *SMEs' management goals*

This theme is defined as the SMEs' senior manager attitudes toward their management goals affecting when they determine their accounting choice. This theme is composed of two subthemes – growth intention and diversity of stakeholders.

7.3.3.1. *Growth intention*

The analysis of current research revealed that SMEs' intention of growing their business is one of the significant determinants when choosing disclosure policy. These subthemes comprise three components: global finance and activities, SMEs' size and Initial Public Offering (IPO) (See Table 7.4).

Subthemes	Interpretative codes	Descriptive codes	Interview ID
Growth intention	Global finance and activities (10 Codes)	There is no case that SMEs in Japan should be supplied finance from financial institutions overseas.	TA4, TA5
		Transactions overseas are usually preoccupied by activities between headquarters in Japan and oversea subsidiaries.	S1, S4
		There is no need to finance from the financial market overseas.	S1, S3, S4
	SMEs' size (25 Codes)	Firm size varies among SMEs.	TA3
		Majority of SMEs are micro-companies.	TA1, TA5, TA3
		There seems to be a tendency that choice of accounting is correlated with SMEs' size.	TA3, TA4, TA5
		Firm's goal is not always to enlarge a company's size.	S2, S3, S4, S5, S7

	Initial public offering (10 Codes)	There are low numbers of SMEs that intend to become public companies.	TA4, TA5
		IPO gives SMEs too much trouble and complexities.	TA4, S5
		Becoming a public company reduces the autonomy of management.	S1, S5
Diversity of stakeholders	Consolidated subsidiary (17 Codes)	Subsidiary SMEs synchronize the accounting standards applied by parent company.	TA3, S3
		CPTAs' work responsibility will be restricted only for taxation when client companies listed in the financial market.	TA5
	SMEs owned by managers and family (10 Codes)	There is little need to apply accounting standards for family-owned SMEs.	S1, TA5,
		SMEs owned by managers have certain discretions when it comes to accounting.	TA3, TA5, S5
	Tax office (24 Codes)	The purpose of SMEs accounting is primarily to maximize tax returns.	TA2, TA5
		For micro-SMEs, taxation is the most influential driver that stipulates accounting practices.	S2, S6, S7
		Earning management is acceptable if it is not legally violated.	TA5
		Tax evasion is not acceptable among CPTAs.	TA1, TA2, TA3, TA4

Table 7.4. *Theme of SMEs' management goal*

One aspect of a firm's growth intention will be related to their willingness in regards to global extension of their business. In general, SMEs expand their global business activities by using finance from global financial market or local banks from overseas. Also, implementing frequent transactions with foreign companies is another way to achieve internationally the goal of business growth. However, participant D of TA4 demonstrated the difficulties for Japanese SMEs to increase their overseas activities and finance as follows:

> "Recently [Japanese SMEs] have attempted to open new factories and/or enter the Chinese market. I got involved in work to prepare [or translate financial statements in Japanese]

> into Chinese. Reporting financial statements was necessary to start new transactions as trade credit before renting the factories in China. People [in China] can't read accounting figures in Japanese [on the financial statements], so we tried to convert it into a Chinese version. There was a Japanese auditing firm in China and we asked for their help to do this conversion. Due to their assistance, it was successfully achieved and we gained what we wanted but financing overseas is beyond my imagination and seems impossible".

Most of the managers from our group of participants admitted that transferring money from Japan would be sufficient to set up overseas subsidiaries or branches for Japanese SMEs (participants of F, I, K, L and P). Thus, this component does not seem to be an effective driver to determine one's accounting standards choice.

The second component in this subtheme is related to SMEs' size. With regard to this aspect, CPTAs in the interviews articulated that SME size has a strong correlation with their choice of accounting policy. They tend to perceive that the larger the company, the more complex accounting standards would need to be applied. In contrast, other top manager participants proposed that there is no such correlation between firm size and accounting standard choice. It is found that even large SMEs ignored adopting SMEs accounting standards and instead applied Tax Accounting Standards for tax return purposes only (e.g. S1). On the contrary, other SMEs comply strictly with J-GAAP, although they are smaller or even micro-SMEs (e.g. S2 and S6). Their choice of accounting standards is subject to factors other than the firm size.

One key component other than firm size was an SMEs' intention of applying for an IPO. Interviewees of CPTAs in this study admitted that SMEs preparing for an IPO would self-impose the adoptation of more complex accounting standards including ASBJ Guideline and J-GAAP. In reality, however, the current research also discovered that only a very small number of SMEs intend to become a public company because they do not want to lose their simple structures and business autonomy, which would occur if they became listed via an IPO (e.g. S1, S5). Participant D's account in TA4 justified this view as follows:

> "[Some of my clients] are qualified to become a public company, but they won't. Surprisingly, there are so many SMEs

that won't go for IPO, especially in XXX [location of the interviewee]. I sometimes suggested my clients go for IPO because they were eligible. When I told them perhaps they need CPAs as the experts or advisors in their company accounting in addition to myself [a CPTA] most of them declined and said "Oh that is not what we would like to do in our company".

7.3.3.2. *Diversity of stakeholders*

Another important subtheme regarding SME management goals was conceptualized as the diversity of stakeholders (Table 7.4). This subtheme comprised three components emerging from the data – consolidated subsidiary, SMEs owned by managers and family, and tax office. This diversity of stakeholders affected SMEs' decision to implement voluntary disclosure in their financial reporting.

In this subtheme, first the SMEs that were consolidated subsidiaries were found to be required to synchronize their accounting standards with those applied by the parent company. In this case, the SMEs' had little discretion in the choice of their disclosure policy. This trait was illustrated by participants' responses regarding accounting behaviors in a 100% consolidated subsidiary. In interview S3, I and K stated:

> "Basically, my company is complying with the accounting standards applied by XXX [parent company name]. There are however some exceptions when applying different accounting standards for revenue and sales that are not used by the parent company, and it is simply because of the specific traits of the industry where my company belongs. [...] when new accounting standards start to be applied [by parent company], the parent company informs [my company] their changes in advance, and they often ask our opinions via questionnaire. The parent company shows respect to our autonomy, and we can negotiate the most efficient way to proceed from these conversations with our parent company".

The company where I and K are employed has applied J-GAAP and was preparing transition toward IFRS due to the influence from the listed parent company, which choice of accounting standards is rare in SMEs.

Second, the majority of SMEs do not separate management from owners; the business is owned by managers and/or family. This causes little or no necessity to prepare financial reports according to specific accounting standards. Because of the lack of demand for financial statements, such SMEs are reluctant to apply SME accounting standards. Participant E of TA5 in this research described this situation as follows:

> "[...] SMEs [owned by managers] have no stakeholders to show their income statement to. I think I will help prepare [income statement] it if it is necessary to be reported [to someone], but I think it is not necessary. [...] CPTAs are generally not willing to suggest that SMEs comply with accrual accounting standards. This is because the SMEs are owned by the managers. I would say such companies [owned by managers] may be able to do earnings management on the financial statements. There are two reasons for this – to increase tax refund and also to make it easier for SMEs to borrow money from banks".

This component is associated with the discussions in the other subthemes and components such as financing needs. If SMEs received loans from banks, the banks would be the primary stakeholder and their pressure may affect a firm's choice of accounting standards. On the other hand, some companies that do not have financial burdens from banks are free to choose their own accounting standards according to their own needs. SMEs of S1 that have no urgent loan requirement from banks selected mandatory Tax Accounting Standard because they are a family-owned company and have no other influential stakeholders. The firm interviewed in S5 also does not owe a debt to their bank and is owned by limited managers. Although they are devoted to improving their accounting system for managerial purposes, they simply deal with lodgment of their tax returns in the first instance.

It was also found that the Tax Office is one of the influential components affecting SMEs' choice of accounting standards. Senior managers, particularly from micro-SMEs, believe that taxation is one of the influential drivers when implementing accounting practices (S2, S6, S7) because applying accounting strategies effectively enables SMEs to maximize their tax returns. In this research, all CPTA participants agreed that tax evasion is not acceptable, but some of them perceived that earnings management is acceptable in order to maximize their tax return. For instance, Participant E's account in TA5 clearly explained the impact of this driver upon the choice of accounting standards.

> "Because I mainly deal with taxation issues, earnings management is acceptable if it does not legally violate our Corporate Tax Act. Clients often ask me to do this. They believe that the best way to deal with earnings is to report a little profit in the bottom line, don't they? Top managers of SMEs don't want to pay any more tax than is legally required. To achieve this, I try to adjust several aspects [of earnings] around two months before the end of the year so as not to report a larger than necessary profit. That's why I [and my clients] don't focus on reporting adequate financial performance through financial reports".

In contrast, it is also true that there are other types of participants who contend that even earnings management is not acceptable. Their focus is not only on reducing tax liabilities as much as possible, but also obtaining loans from lenders more easily. Participant A's account in TA1 presented this view when he was asked whether or not he would apply a regular periodic depreciation that is permitted by General Standards.

> "If avoiding regular periodic depreciation is not for tax evasion, then it is possible to apply it. But even in this case, such a procedure causes problems in terms of obtaining finance from the banks. So it is very obscure to judge [if this is good or bad] […] Not to apply regular periodic depreciation does not make sense to me because the banks understand this. So, if the clients want to apply this technique, it is not my responsibility but rather the firm's. It is possible to be applied simply because its procedure is acceptable for taxation".

As seen, the impact of this component upon SMEs' choice of accounting standards can be subject to their individual financial needs. Their actual choice of standards will be determined by balancing taxation and financial needs.

7.4. Discussion and interpretation

The analysis of the interview data comprised the following recurring issues regarding voluntary disclosure for SMEs: a trading off between expertise and autonomy; accrual versus tax accounting; checklist scheme and growth intention versus maintaining the status quo.

7.4.1. *Trading off between expertise and autonomy*

Whether or not SMEs rely on external accountants for their accounting practices was found to be the key factor to drive their voluntary disclosure. This result was consistent with prior literature that contends SMEs demand business advice from external accountants because of deficiencies in their in-house expertise and resources [BEN 05, CHR 04, GOO 04, COL 02, ROB 00]. In contrast, such deficiencies in expertise and resources were not comprised in the subtheme of accounting practice independent from professionals, because SMEs who are independent from the profession must have strong confidence in dealing with accounting by themselves. Smaller SMEs with resource constraints in accounting also enabled them to establish high-quality in-house accounting by using information technologies. This is confirmed by Blackburn and Jarvis [BLA 10] who state that extensive use of computer technologies has been applied recently to manage compliance services of SMEs accountancy practices.

One example of this is the application of cloud accounting technology for SMEs. According to Mongan [MON 11], cloud computing technology refers to the centralization of all or part of a firm's computer resources via a shared provider of such services. This technology has been recently used in accounting information systems [BRA 15, GUP 13]. The present research revealed evidence that information technology reinforces the weaknesses of smaller SMEs to work as the substitute for an in-house financial department that large firms normally facilitate. Extant research also affirmed that cloud technology reduces the gap between SMEs and large organizations since it requires lower investment in terms of equipment and licenses [QUI 14, ION 13]. Given this technology support, even smaller SMEs can maintain autonomy in accounting. Such an autonomy would drive firms to apply accounting for strategic management purposes in addition to financial purposes. This autonomy is thought to be of value in an attempt to obtain sustainable development of the SMEs regardless of their small organization size.

7.4.2. *Cloud accounting for value creation*

Several prior studies pointed out that using cloud technologies in the accounting information system has a positive impact by significantly reducing acquisition cost, maintenance and management of the company (e.g. [BRA 15, ION 13]). This obviously has the effect of gaining economic value for the companies by applying cloud accounting. There are several studies that investigate how the use of cloud accounting infrastructure affected the business performance of SMEs (e.g. [CLE 16, QUI 14]). Eventually, Cleary and Quinn [CLE 16] reported empirical evidence that cloud accounting had a positive and statistically significant impact on human capital and related capital. The other study of Ionescu and Prichici [ION 13] articulated how the replacement process of cloud accounting systems reduced the need for specialist skills and allowed managements to focus on the most important aspects of a business. This process is interpreted to not generate a gain in the short term but in the long run SMEs may be able to develop their competitive advantage in making use of spare time saved by adopting to the cloud system.

7.4.3. *Accrual versus tax accounting*

The two subthemes of the professionals' attitude and the diversity of stakeholders address the argument about a choice between accrual and tax accounting. The reason why these two dichotomous constructs emerged among participants could be because of the close connection between financial and tax accounting for SMEs in Japan. Previous literature argued that the local taxation system is one of the main institutional factors that would have a significant influence on a national accounting system [WEH 14]. Eventually large studies found such a close connection in several countries including Germany [WEH 14, HAL 12, HAL 92], Austria [EBE 07], Spain [GAL 04] and France [LAM 98]. Japan is also categorized in this group where Japanese accounting standards for SMEs accord with the Corporate Tax Act, therefore their current net income is their taxable income [FUJ 15]. This structure of the standard causes the "principle of congruency", which is thought to enhance SMEs managers to prepare financial statements in conformity with the accounting rules in the Corporation Tax Act [FUJ 15].

Given this principle of congruency, earlier studies empirically found that manager-controlled firms adopted an accounting method that enabled them to increase their tax returns [HOL 83, WAT 86]. However, recent research by Fujibayashi, Kojima and Tsuji [FUJ 15] examined family-controlled SMEs in Japan and revealed that good performing SMEs chose earnings management that allowed their accounting profit to increase their tax returns, but they did not decrease their profit to the degree to which the earnings management might have caused loss of trust with their lenders.

Compared with previous empirical studies, the research of this chapter affirmed that SMEs' choice between accrual and tax accounting was determined based on whether they required finance or not and whether their primary stakeholder was a bank or the tax office. In addition, the contribution of this research was to reveal that this choice would be highly influenced in relation to selecting accounting professionals as the firms' advisors. It seems that SMEs' accounting behaviors are driven not to reduce the cost of capital and enhance signals of quality but rather from suggestions and advice from the firm's reliable advisor. Their suggestions also help dictate which standards will be applied by SMEs – ASBJ Guidelines or General Standards.

7.4.4. *Checklist scheme*

The checklist scheme was also found to be the key determinant for choice of accounting standard among SMEs particularly those that need finance. The SMEs in this category are led to apply either ASBJ Guidelines or General Standards simply to reduce their cost of capital, which would at least lead to obtaining economic value for SMEs. However, this behavior is not likely to be explained by agency theory because SMEs' choice of accounting policy is regulated by the checklist schemes that are already enacted as the formal system by FSA and SMEA. Therefore, their decision of disclosure with this scheme will be made regardless of information asymmetry between firms and banks. The submission of this checklist may result in reducing information asymmetry but this scheme does not seem to work effectively because it was reported in our interviews that bankers lack sufficient knowledge about the scheme, which causes serious distortion when choosing the accounting standards for SMEs.

7.4.5. *Growth intentions versus preserving status quo*

Prior studies indicated that SMEs' choice of accounting standards providing global comparable information is significantly associated with their cross-border activities such as finance from foreign investors and lenders [EIE 13, AND 11] and foreign exports [KIL 14, UYA 13, EIE 13]. However, our interviews demonstrated such an international growth intention is scarce among Japanese SMEs. Thus, this component is not likely to positively influence their choice of standards. In Japan, ASBJ Guidelines are supposed to be the global comparable financial standard that integrates with the IFRS framework [KAW 12], but this characteristic was not clearly identified among SME managers in the current research because of their low demand for finance by international capital providers and lesser need for trade credit by counterparts overseas. Prior studies reported that international reporting strategies using International Accounting Standards or US GAAP directly improved the economic value of the companies (e.g. [LEU 00]). However, the present study showed inconsistent results that the CPTAs and SME managers interviewed in this research perceived the strategy of global voluntary disclosure not to be economically valued for them. This finding was also supported by the deductive statistical test implemented in Chapter 6.

With regard to company size, most of the CPTAs participants agreed that larger SMEs tend to apply more complicated standards than smaller SMEs, while this association was not perceived by the SME managers interviewed in this research. Extensive studies have examined empirically the effect of a firm's size as a predictor for voluntary choice of accounting standards [EIE 13, ALB 13b, AND 12, LIT 12, QUA 12, CAR 10, EIE 09], but these outcomes vary in each study and do not reach a solid consensus. Among them, the finding of this chapter was consistent with Eierle and Haller [EIE 09], where a firm's size is not related with their demand for voluntary disclosure of financial statements.

Growth intention somehow represents a firm's ambitions to transition from a private to a public company. In the accounting literature, voluntary adaptation of global standards has been seen especially in emerging economies where some SMEs would apply IFRS in order to make their transition to full IFRS easier [KIL 14, UYA 13, MUL 10]. Similarly, it was found that our CPTA participants agreed with the idea that Japanese SMEs who intend to do IPO would adopt more complex accounting standards such as ASBJ Guidelines or J-GAAP. However, Japanese SMEs in the present research were also found not to use external equity for their finance needs very often. The participant's accounts in this research reported that owners of SMEs, where their business was qualified to become a public company, feared loss of autonomy and decided to preserve their status quo. Prior studies support this finding by applying the pecking order theory where decisions are influenced by the owner's desire to maintain control [BER 00, CRE 95].

As a result of the above interpretations, growth intention is theoretically thought to be an influential subtheme for voluntary disclosure, although SMEs in Japan examined by this research are not strongly affected by this subtheme.

7.5. Conclusion

This chapter aimed to identify the factors that determine SME choice of financial accounting standards for implementing voluntary disclosure in Japan. For this purpose, the chapter used a thematic analysis method using interview data collected from CPTAs and senior managers from SMEs in Japan. The findings of this study were as follows :

1) support for accounting practices provided by an external accountant (as an advisor) offset the SMEs' autonomy of accounting. The choice of accounting policy in this case would be driven by their external accountants' suggestions in accordance with their attitude toward accrual accounting. It was also found that small and less resourced SMEs tend to lose their discretion for accounting policy choice but the adoption of information technologies such as a cloud accounting system helped them maintain accounting autonomy. This autonomy is regarded as producing long-term value for SMEs' sustainable development;

2) as far as the SMEs were not independent from banks and accounting experts, the majority of SME managers in this research preferred to choose accrual accounting standards for finance purposes. Professional advisors from external sources would also strongly influence which standards SMEs chose;

3) a checklist scheme was one of the seminal drivers to determine SMEs' choice of accounting standards, but it was found that banks' lack of knowledge and their incorrect perceptions regarding this scheme were distorting an adequate choice of accounting standards;

4) international growth intention was found to be scarce among Japanese SMEs, thus their motivation for adopting comparable global reporting standards was very low. This finding explained the reason why the ASBJ Guidelines and IFRS for SMEs are hardly chosen in SME accounting standards;

5) the present research confirmed that firm size was not related to the demand for the voluntary disclosure of financial statements;

6) there is no intention of using ASBJ Guidelines and J-GAAP for transition purposes among SMEs, since the majority of SMEs tended to preserve the status quo in terms of business growth and did not have high motivation to become a listed company.

Giving these findings, this research shed light on the recurring factors that influence SMEs' voluntary disclosure. Extant theories such as agency theory, signaling theory and proprietary cost theory could be typically used to explain the mechanism of accounting policy choice of voluntary disclosure of financial statements for large public companies, but these theories were found not to be suitable for SMEs. The choice of accounting standards among SMEs is neither driven by firms' intentions to reduce capital costs nor to increase corporate values. The key to addressing firms' value for SMEs is attributed to autonomy in accounting.

In this study, a thematic analysis method was applied to conduct an exploratory study, which successfully identified inductive relations and patterns between seminal influential drivers from which new theories may be developed. In particular, the present research revealed unintended findings that SMEs' choice of accounting policy will be highly influenced by external accountants' professional attitude regarding financial accounting. Future research could focus on some aspects relating to this attitude such as professional education and training.

Further, this study contributes to the accounting literature in the sense that the findings are potentially relevant to policy. In this research, a checklist scheme was found to be harmful from an adequate financial accounting perspective. This result is useful for standard setters and policy makers in assessing the impact of regulations and policies on SMEs.

There are some limitations to the current study that need to be considered. One primary limitation is that we did not collect the data from the SMEs that applied ASBJ Guidelines. Although this research obtained information and comments about ASBJ Guidelines from CPTAs who have clients that apply this standard, the absence of a firm adopting ASBJ Guidelines was a significant omission. This flaw should be addressed in future studies. Similarly, data should be collected from other important stakeholders such as banks and authorities to improve the quality of this research.

Even considering the limitations, the findings of the study provide in-depth insight beyond the previous literature, which has usually been underpinned by quantitative research. This insight is important considering the similar settings of SMEs in other countries and future opportunities for international comparative studies.

Bibliography

[ABE 96] ABERNETHY M.A., CHUA W.F., "A field study of control systems 'redesign': the impact of institutional processes on strategic choice", *Contemporary Accounting Research*, vol. 13, no. 2, pp. 569–606, 1996.

[ACC 13] ACCA "Technology trends: their impact on the global accountancy profession", available at: http://www.accaglobal.com/content/dam/acca/ global/PDF-technical/futures/pol-af-ttti.pdf, 2013.

[ADL 96] ADLER P., BORYS B., "Two types of bureaucracy: enabling and coercive", *Administrative Science Quarterly*, vol. 41, no. 1, pp. 61–90, 1996.

[AFU 00] AFUAH A., TUCCI C.L., *Internet Business Models and Strategies: Text and Cases*, Irwin/McGraw-Hill, New York, 2000.

[AGY 16] AGYEU-MENSAH B.K., "Internal control information disclosure and corporate governance: evidence from an emerging market", *Corporate Governance*, vol. 16 no. 1, pp. 79–95, 2016.

[AHM 99] AHMED K., CORTIS J., "Associations between corporate characteristics and disclosure levels in annual reports: a meta-analysis", *British Accounting Review*, vol. 31, no. 1, pp. 35–61, 1999.

[AHR 04] AHRENS T., CHAPMAN C.S., "Accounting for flexibility and efficiency: a field study of management control systems in a restaurant chain", *Contemporary Accounting Research*, vol. 21, no. 2, pp. 271–301, 2004.

[ALB 12] ALBU N., ALBU C.N., "Strategies for and implications of the IFRS for SMEs implementation in emerging economies", *Comptabilités et innovations*, 2012.

[ALB 13a] ALBU C.N., ALBU N., PALI-PISTA S.F. *et al.*, "Implementation of IFRS for SMEs in emerging economies: stakeholder perceptions in the Czech Republic, Hungary, Romania and Turkey", *Journal of International Financial Management and Accounting*, vol. 24, no. 2, pp. 141–175, 2013.

[ALB 13b] ALBU C.N., ALBU N., GÎRBINĂ M.M., "Accounting for SMEs – how important is size in choosing between global and local standards?", *International Journal of Social, Behavioral, Educational, Economic and Management Engineering*, vol. 7, no. 4, pp. 441–445, 2013.

[ALC 65] ALCHIAN A., "Some economics of property rights", *Il Politico*, vol. 30, pp. 816–829, 1965.

[ALC 72] ALCHIAN A., DEMSETZ H., "Production, information costs and economic organization", *American Economic Review*, vol. 62, no. 5, pp. 777–795, 1972.

[ALC 73] ALCHIAN A., DEMSETZ H., "The property right paradigm", *Journal of Economic History*, vol. 33, no. 1, pp. 16–27, 1973.

[ALK 06] ALKARAAN R. NORTHCOTT D., "Strategic capital investment decision-making: a role for emergent analysis tools? A study of practice in large UK manufacturing companies", *The British Accounting Review*, vol. 38, pp. 149–173, 2006.

[ALL 09] ALLEE K.D., YOHN T.L., "The demand for financial statements in an unregulated environment: an examination of the production and use of financial statements by privately held small businesses", *The Accounting Review*, vol. 84, no. 1, pp. 1–25, 2009.

[ALO 87] ALONSO P., "Introduction", in ALONSO P., STARR W. (eds), *The Politics of Numbers*, Russell Sage Foundation, New York, 1987.

[AMI 01] AMIT R., ZOTT C., "Value Creation in e-Business", *Strategic Management Journal*, vol. 22, no. 6/7, pp. 493–520, 2001.

[AND 12] ANDRÉ P., WALTON P.J., YANG D., "Voluntary adoption of IFRS: a study of determinants for UK unlisted firms", available at: SSRN1978986, 2012.

[ANT 65] ANTHONY R.N., Planning and Control Systems: A Framework for Analysis, Division of Research, Graduate School of Business Administration, Harvard University, Boston, 1965.

[APP 01] APPLEGATE L., "E-business models: Making sense of the Internet business landscape", in DICKSON G., DESANCTES G. (eds), *Information Technology and the Future Enterprise: New Models for Managers*, Prentice Hall, Upper Saddle River, 2001.

[ARR 16] ARRAOU P., "The certified public accountant and the digital economy", Report including excerpts from the Introduction to the Opening Plenary, 70th Congress of the Order of Certified Public Accountants, 2015, available at: http://ceccar.ro/ro/wp-content/uploads/2016/10/expert-comptable-economie-numerique_ENG.pdf, 2016.

[ARS 08] ARSLAN O., ER I.D., "SWOT analysis for safer carriage of bulk liquid chemicals in tankers", *Journal of Hazardous Materials*, vol. 154, no. 1, pp. 901–913, 2008.

[ASH 56] ASHBY, W.R., *An Introduction to Cybernetics*, John Wiley, New York, 1956.

[ATI 10] ATIK A. "SME's views on the adoption and application of "IFRS for SMEs" in Turkey", *European Research Studies*, vol. 13 no. 4, pp. 19–32, 2010.

[BAR 91] BARNEY J.B., "Firm resources and sustained competitive advantage", *Journal of Management*, vol. 17, no. 1, pp. 99–120, 1991.

[BAR 95] BARNEY J.B., "Looking inside for competitive advantage", *The Academy of Management Executive*, vol. 9, no. 4, pp. 49–61, 1995.

[BAS 12] BASSEMIR M., "Why do private firms adopt IFRS?", available at: SSRN 1896283, 2012.

[BAT 01] BATALHA M.O., *Gestão agroindustrial*, Atlas, São Paulo, 2001.

[BAY 14] BAY C., CATASUS B., JOHED G., "Situating financial literacy", *Critical Perspective on Accounting*, vol. 25, no. 1, pp. 36–45, 2014.

[BEN 05] BENNETT R.J., ROBSON P.J.A., "The advisor-SME client relationship: impact, satisfaction and commitment", *Small Business Economics*, vol. 25, no. 3, pp. 255–271, 2005.

[BER 00] BERGGREN B., OLOFSSON C., SILVER L., "Control aversion and the search for external financing in Swedish SMEs", *Small Business Economics*, vol. 15, no. 3, pp. 233–242, 2000.

[BER 16] BERNARD T.S., "Tax tips for those who make money in the gig economy", *The New York Times*, available at: http://www.nytimes.com/2016/ 03/05/your-money/tax-tips-for-those-who-make-money-in-the-gig-economy.html?_r=0, 2016.

[BHI 03] BHIMANI A., *Management Accounting in the Digital Economy*, Oxford University Press, Oxford, 2003.

[BHI 09] BHIMANI A., BROMWITCH M., "Management accounting in a digital and global economy: the interface of strategy, technology, and cost information", in CHAPMAN C.S., COOPER D.J., MILLER P. (eds), *Accounting, Organizations, and Institutions: Essays in Honour of Anthony Hopwood*, Oxford University Press, Oxford, 2009.

[BHI 14] BHIMANI A., WILLCOKS L., "Digitization, 'Big data', and the transformation of accounting information", *Accounting and Business Research*, vol. 44, no. 4, pp. 469–490, 2014.

[BLA 10] BLACKBURN R., JARVIS R., "The role of small and medium practices in providing business support to small- and medium-sized enterprises", *Information Paper, Small- and Medium Practices Committee, International Federation of Accountants*, 2010.

[BLA 96] BLANCHARD K.H., CARLOS J.P., RANDOLPH W.A., *Empowerment Takes More Than a Minute*, Berrett-Koehler Publishers, Oakland, 1996.

[BOE 07] BOESSO G., KUMAR K., "Drivers of corporate voluntary disclosure: a framework and empirical evidence from Italy and the United States, accounting", *Auditing & Accountability Journal*, vol. 20, no. 2, pp. 269–296, 2007.

[BOY 98] BOYATZIS R.E., *Transforming Qualitative Information: Thematic Analysis and Code Development*, Sage Publications, London, 1998.

[BRA 06] BRAUN V., CLARKE V., "Using thematic analysis in psychology", *Qualitative Research in Psychology*", vol. 3, no. 2, pp. 77–101, 2006.

[BRA 15] BRANDAS C., MEGAN O., DIGRAGA O., "Global perspectives on accounting information systems: mobile and could approach", *Procedia Economics and Finance*, vol. 20, pp. 88–93, 2015.

[BRE 88] BREALEY R.A., MYERS S.C., *Principles of Corporate Finance,* 9th ed., McGraw Hill Higher Education, New York, 1988.

[BRO 11] BROWN P., "International financial reporting standards: what are the benefits?", *Accounting and Business Research*, vol. 41, no. 3, pp. 269–285, 2011.

[BRO 85] BROWNELL P., "Budgetary systems and the control of functionally differentiated organizational activities", *Journal of Accounting Research*, vol. 23, no. 2, pp. 502–512, 1985.

[BUL 16] BULOW (VON) REVEL T., BOLARDI P., "The 'Uber-economy'", Clipperton Finance, available at: http://www.clipperton.net/newsletter/The%20Uber%20Economy%20-%20Clipperton%20paper%20-%20vf.pdf, 2016.

[BUN 12] BUNEA S., SĂCĂRIN M., MINU M., "Romanian professional accountants' perception on the differential financial reporting for small and medium-sized enterprises", *Accounting and Management Information Systems*, vol. 11, no. 1, pp. 27–43, 2012.

[BUS 13] BUSINESS ACCOUNTING COUNCIL, *An Interim Policy Relating to International Accounting Standards*, BAC, available at: http://www.fsa.go.jp/news/24/sonota/20130620-2/01.pdf, 2013.

[BUT 07] BUTLER M., FRAFT A., WEISS I.S., "The effect of reporting frequency on the timeliness of earnings: the cases of voluntary and mandatory interim reports", *Journal of Accounting and Economics*, vol. 43, nos. 2–3, pp. 181–217, 2007.

[CAR 10] CARFANG T.A., "IFRS and SMEs", *Accounting Today*, October 25, 2010.

[CAR 14] CARPENTER M., DAIDJ N., MORENO C., "Game console manufacturers: the end of sustainable competitive advantage?", *Digiworld Economic Journal*, vol. 1, no. 94, pp. 39–60, 2014.

[CAS 09] CASADESUS-MASANELL R., RICART J.E., "From strategy to business models and to tactics", Working Paper, 10-036, Harvard Business School, available at: http://www.hbs.edu/faculty/Publication%20Files/10-036.pdf, 2009.

[CAS 10] CASADESUS-MASANELL R., RICART J.E., "From strategy to business models and to tactics", *Long Range Planning*, vol. 43, no. 2, pp. 195–215, 2010.

[CAS 15] CASSAR G., CHRISTOPHER D.I., CAVALLUZZO K.S., "Alternative information sources and information asymmetry reduction: evidence from small business debt", *Journal of Accounting and Economics*, vol. 59, nos. 2–3, pp. 242–263, 2015.

[CEU 17] CEUSTERMANS S., BREESCH D., "Determinants of voluntary disclosure of sales in small private companies in Belgium", *Journal of International Financial Management and Accounting*, vol. 28, no. 2, pp. 172–204, 2017.

[CHA 05] CHAPMAN C., *Controlling Strategy: Management, Accounting, and Performance Measurement*, Oxford University Press, New York, 2005.

[CHA 09] CHAPMAN C., COOPER D.J., MILLER P., *Accounting, Organizations, and Institutions: Essays in Honour of Anthony Hopwood*, Oxford University Press, Oxford, 2009.

[CHE 02] CHESBROUGH H., ROSEMBLOOM R., "The role of business models in capturing value from innovation: evidence from Xerox Corporation's technology spin-off companies" *Industrial and Corporate Change*, vol. 11, no. 3, pp. 529–555, 2002.

[CHE 05] CHENHALL R.H., "Content and process approaches to studying strategy and management control systems", in CHAPMAN C.S. (ed.), *Controlling Strategy: Management Accounting and Performance Measurement*, pp. 10–36, Oxford University Press, New York, 2005.

[CHE 07] CHESBROUGH H., "Business model innovation: it's not just about technology anymore", *Strategy & Leadership*, vol. 35, no. 6, pp. 12–17, 2007.

[CHR 05] CHRISMAN J.J., MAMULLAN E., HALL J., "The influence of guided preparation on the long-term performance of new ventures", *Journal of Business Venturing*, vol. 20, no. 6, pp. 769–791, 2005.

[CIA 13] CIASULLO M.V., TROISI O., "Sustainable value creation in SMEs: a case study", *The TQM Journal*, vol. 25, no. 1, pp. 44–61, 2013.

[CLE 16] CLEARY P., QUINN M., "Intellectual capital and business performance: an exploratory study of the impact of cloud-based accounting and finance infrastructure", *Journal of Intellectual Capital*, vol. 17, no. 2, pp. 255–278, 2016.

[COL 02] COLLIS J., JARVIS R., "Financial information and the management of small private companies", *Journal of Small Business and Enterprise Development*, vol. 9, no. 2, pp. 100–110, 2002.

[CON 11] CONNELLY B.L., CERTO S.T., IRELAND R.D. *et al.*, "Signalling theory: a review and assessment", *Journal of Management*, vol. 37, no. 1, pp. 139–167, 2011.

[COO 94] COOPER R., "Kyocera corporation: the amoeba management system", *Harvard Business School Case* (9-195-064), 1994.

[COO 95] COOPER R., *When Lean Enterprises Collide: Competing Through Confrontation*, Harvard Business School Press, Boston, 1995.

[COP 06] COPPENS C., VAN WYMEERSCH K., ENGELS L., "The attitude of Belgian SMEs towards the implementation of IAS/IFRS: an inquiry into what Belgian SMEs know and expect, and how well they are prepared for IAS/IFRS", *Annual Congress of the European Accounting Association*, Dublin, Ireland, 2006.

[CRE 95] CRESSY R., "Business borrowing and control: a theory of entrepreneurial types", *Small Business Economics*, vol. 7, no. 4, pp. 291–300, 1995.

[DAI 10] DAIDJ N., GRAZIA C., HAMMOUDI A., "Introduction to the non-cooperative approach to forming coalitions: the case of the Blu-Ray/HD-DVD standards' war", *Journal of Media Economics*, vol. 23, no. 4, pp. 192–215, 2010.

[DAI 15] DAIDJ N., *Developing Strategic Business Models and Competitive Advantage in the Digital Sector*, IGI Global, Hershey, 2015.

[DAI 17] DAIDJ N., "Uberization (or uberification) of the economy", *Encyclopedia of Information Science and Technology*, 4th ed., Hershey, 2017.

[DAV 94] D'AVENI R.A, *Hypercompetition: Managing the Dynamics of Strategic Manoeuvring*, Free Press, New York, 1994.

[DAV 10] D'AVENI R.A., DAGNINO G.B., SMITH K.G., "The age of temporary advantage", *Strategic Management Journal*, vol. 31, no. 13, pp. 1371–1385, 2010.

[DAV 06] DAVENPORT T.H., "Competing on Analytics", *Harvard Business Review*, vol. 84, no. 1, pp. 98–107, 2006.

[DAV 05] DAVILA T., "The promise of management control systems for innovation and strategic change", in CHAPMAN C.S. (ed.), *Controlling Strategy: Management Accounting and Performance Measurement*, Oxford University Press, New York, 2005.

[DEE 00] DEEGAN C., *Financial Accounting Theory*, McGraw-Hill, Sydney, 2000.

[DEM 67] DEMSETZ H., "Towards a theory of property rights", *American Economic Review*, vol. 57, pp. 347–373, 1967.

[DEP 00] DEPOERS F., "A cost benefit study of voluntary disclosure: some empirical evidence from French listed companies", *European Accounting Review*, vol. 9, no. 2, pp. 245–263, 2000.

[DEV 01] DE VILLIERS C., LUBBE D.S., "Industry differences in respect of corporate environmental reporting in South Africa: a research note", *Meditari Accountancy Research*, vol. 9, no. 1, pp. 81–91, 2001.

[DEW 99] DE WIT B., MEYER R., *Strategy Synthesis*, Thompson, London, 1999.

[DOR 15] DORNER K., EDELMAN D., *What 'digital' really means*, Mckinsey & Company: Insights and Publications, 2015.

[EBE 07] EBERHARTINGER E., KLOSTERMANN M., "What If IFRS were a tax base? New empirical evidence from an Austrian perspective", *Accounting in Europe*, vol. 4, no. 2, pp. 141–168, 2007.

[EIE 09] EIERLE B., HALLRE A., "Does size influence the suitability of the IFRS for small and medium-sized entities? – empirical evidence from Germany", *Accounting in Europe*, vol. 6, no. 2, pp. 195–230, 2009.

[EIE 13a] EIERLE B., HELDUSER C., What drives SMEs need to provide internationally comparable accounting information? Empirical evidence from Germany, Working Paper, University of Bamberg, 2013.

[EIE 13b] EIERLE B., SCHULTZE W., "The role of management as a user of accounting information: implications for standard setting", *Journal of Accounting and Management Information Systems*, available at SSRN: https://ssrn.com/abstract=1130162, 2013.

[EIS 02] EISENMANN T.R., *Internet Business Models*, McGraw Hill, New York, 2002.

[EZZ 04] EZZAMEL M., SIMON L., WILLMOTT H., "Accounting representation and the road to commercial salvation", *Accounting, Organizations and Society*, vol. 29, pp. 783–813, 2004.

[FIG 02] FIGGE F., HAHN T., SCHALTEGGER S. *et al.*, "The sustainability balance scorecard-linking sustainability management to business strategy", *Business Strategy and the Environment*, vol. 11, no. 5, pp. 269–284, 2002.

[FLA 83] FLAMHOLZ E.G., "Accounting, budgeting and control systems in their organizational context: theoretical and empirical perspectives", *Accounting, Organizations and Society*, vol. 8, no. 2, pp. 153–169, 1983.

[FRA 08] FRANCIS J.R., KHURANA I.K., MARTIN X. *et al.*, "The Role of Firm-Specific Incentives and Country Factors in Explaining Voluntary IAS Adoption: Evidence from Private Firms", *European Accounting Review*, vol. 17, no. 2, pp. 331–360, 2008.

[FRO 10] FROW N., MARGINSON D., OGDEN S., "Continuous budgeting: reconciling budget flexibility with budgetary control", *Accounting, Organizations and Society*, vol. 35, no. 4, pp. 444–461, 2010.

[FUJ 15] FUJIBAYASHI M., KOJIMA H., TSUJI M., "Financial reporting quality of family-controlled small and medium-sized entities: a Japanese case study", *Academy of Accounting and Financial Studies Journal*, vol. 19, no. 3, pp. 87–102, 2015.

[FUR 72] FURUBOTN E.G., PEJOVITCH S., "Property rights and economic theory: a survey of recent literature", *Journal of Economic Literature*, vol. 10, pp. 1137–1162, 1972.

[GAL 04] GALLEGO I., "The accounting and taxation relationship in Spanish listed firms", *Managerial Auditing Journal*, vol. 19, no. 6, pp. 796–819, 2004.

[GID 71] GIDDENS G.A., *Capitalism and Modern Social Theory*, Cambridge University Press, Cambridge, 1971.

[GOO 04] GOODERHAM P.N., TOBIASSEN A., DOVING E. *et al.*, "Accountants as sources of business advice for small firms", *International Small Business Journal*, vol. 22, no. 1, pp. 5–22, 2004.

[GRA 91] GRANT R.M., "The resource-based theory of competitive advantage: implications for strategy", *California Management Review*, vol. 22, pp. 114–135, 1991.

[GRA 05] GRANT R.M., *Contemporary Strategy Analysis*, Blackwell Publishing, Oxford, 2005.

[GRO 86] GROSSMAN S.J., HART O., "The costs and benefits of ownership: a theory of vertical and lateral integration", *Journal of Political Economy*, vol. 94, no. 4, pp. 691–719, 1986.

[GUB 89] GUBA E.G., LINCOLN Y.S., *Fourth Generation Evaluation*, SAGE Publications, Thousand Oaks, 1989.

[GUE 06] GUEST G., BUNCE A., JOHNSON L., "How many interviews are enough? An experiment with data saturation and variability", *Field Methods*, vol. 18, no. 1, pp. 59–82, 2006.

[GUP 13] GUPTA P., SEETHARAMAN A., RAJ J.R., "The usage and adoption of could computing by small and medium businesses", *International Journal of Information Management*, vol. 33, no. 5, pp. 861–874, 2013.

[HAB 12] HABTAY S.R., "A Firm-Level Analysis on the Relative Difference between Technology-Driven and Market-Driven Disruptive Business Model Innovations", *Creativity and Innovation Management*, vol. 21, no. 3, pp. 290–303, 2012.

[HAL 11] HALL M., "Do comprehensive performance measurement systems help or hinder managers' mental model development?" *Management Accounting Research*, vol. 22, no. 1, pp. 68–83, 2011.

[HAL 92] HALLER A., "The relationship of financial and tax accounting in Germany: a major reason for disharmony in Europe", *The International Journal of Accounting*, vol. 27, no. 4, pp. 310–323, 1992.

[HAL 12] HALLER A., FERSTL E.M., The relationship of financial and tax accounting in Germany – a revisit after 20 years, Working Paper, University of Regensburg, 2012.

[HAM 02] HAMEL G., *Leading the Revolution: How to Thrive in Turbulent Times by Making Innovation a Way of Life*, Harvard Business School Press, Boston, 2002.

[HAN 00] HAN J.C.Y., WILD J., "Predisclosure information, firm capitalization, and earnings information transfers", *Journal of Business Research*, vol. 49, no. 3, pp. 273–288, 2000.

[HAN 05] HANSEN A., MOURITSEN J., "Strategies and organizational problems: constructing corporate value and coherence in balanced scorecard processes", in CHAPMAN C.S. (ed.), *Controlling Strategy: Management Accounting and Performance Measurement*, Oxford University Press, New York, 2005.

[HAR 01] HARGADORN A.B., YELLOWLEES D., "When Innovations Meet Institutions: Edison and the Design of the Electric Light", *Administrative Science Quarterly*, vol. 46, no. 3, pp. 476–501, 2001.

[HAR 90] HART O., MOORE J., "Property rights and the nature of the firm", *Journal of Political Economy*, vol. 98, no. 6, pp. 1119–1158, 1990.

[HAT 15] HATA T., "Introductory Colum, in KCCS Management News, Kyocera Communication Systems Corporation, (in Japanese), available at: https://amc.kccs.co.jp/case_colum_report/mailmagazine/2015/magazine_vol47.html, 2015.

[HAW 01] HAWKINS R., The 'Business Model' as a Research Problem in Electronic Commerce, STAR IST Project, Issue Report 4, 2001.

[HEA 99] HEALY P.M., HUTTON A., PAPEPU K., "Stock performance and intermediation changes surrounding sustained increases in disclosure", *Contemporary Accounting Research*, vol. 16, no. 3, pp. 485–312, 1999.

[HEA 01] HEALY P.M., PALEPU K.G., "Information asymmetry, corporate disclosure, and the capital markets: A review of the empirical disclosure literature", *Journal of Accounting and Economics*, vol. 31, nos. 1–3, pp. 405–440, 2001.

[HEN 06] HENRI J.F., "Management control systems and strategy: a resource-based perspective", *Accounting, Organizations and Society*, vol. 31, no. 6, pp. 529–558, 2006.

[HES 16] HESS T., MATT C., WIESBÖCK F. *et al.* "Options for formulating a digital transformation strategy", *MIS Quarterly Executive*, vol. 15, no. 2, pp. 103–119, 2016.

[HIL 05] HILTON R.W., *Managerial Accounting: Creating Value in a Dynamic Business Environment*, 6th ed., McGraw Hill, Boston, 2005.

[HIR 09] HIROMOTO T., *The Management System for Autonomous Organizations: Wisdom of Japanese Management*, Moriyama-Shoten, Tokyo, 2009.

[HO 13] HO P.L., TAYLOR G., "Corporate governance and different types of voluntary disclosure: evidence from Malaysian listed firms", *Pacific Accounting Review*, vol. 25, no. 1, pp. 4–29, 2013.

[HOL 83] HOLTHAUSEN R.W., LEFTWICH R.W., "The economic consequences of accounting choice: implications of costly contracting and monitoring", *Journal of Accounting and Economics*, vol. 5, no. 2, pp. 77–117, 1983.

[HOP 03] HOPE J., FRASER R., *Beyond Budgeting: How Managers Can Break Free from the Annual Performance Trap*, Harvard Business School Press, Boston, 2003.

[HOP 13] HOPE O.K., THOMAS W.B., VYAS D., "Financial reporting quality of US private and public firms", *The Accounting Review*, vol. 88, no. 5, pp. 1715–1742, 2013.

[HOP 17] HOPE O.K., THOMAS W.B., VYAS D., "Stakeholder demand for accounting quality and economic usefulness of accounting in U.S. private firms", *Journal of Accounting and Public Policy*, vol. 36, no. 1, pp. 1–13, 2017.

[HOP 72] HOPWOOD A.G., "An empirical study of the role of accounting data in performance evaluation", *Journal of Accounting Research*, vol. 10, no. 1, pp. 156–182, 1972.

[HOR 09] HORII S., "Rationality of payback period method: implication from a case study", *Ritsumeikan Business Journal*, vol. 47, no. 6, pp. 53–67, 2009.

[IAN 04] IANSITI M., LEVIEN R., *The Keystone Advantage: What the New Dynamics of Business Ecosystems Mean for Strategy, Innovation, and Sustainability*, Harward Business School Press, Boston, 2004.

[INA 01] INAMORI K., *Passion for Success: Passion*, PHP Laboratory, Tokyo, (In Japanese), 2001.

[INA 13] INAMORI K., *Amoeba Management: The Dynamic Management System for Rapid Market Response*, CRC Press, New York, 2013.

[INC 97] INCHAUSTI B.G., "The influence of company characteristics and accounting regulation on information disclosed by Spanish firms", *European Accounting Review*, vol. 6, no. 1, pp. 45–68, 1997.

[INT 09] INTERNATIONAL ACCOUNTING STANDARD BOARD, International Financial Reporting Standard for Small and Medium-Sized Entities, Report, 2009.

[ION 13] IONESCU B.S., PRICHICI C., "Potential beneficiaries of cloud accounting technology: small or large companies?", *Change and Leadership*, vol. 17, pp. 282–292, 2013.

[JAC 89] JACQUEMIN A., BUIGUES P., ILZKOVITZ F., "Horizontal mergers and competition policy in the European Community", *European Economy*, vol. 40, pp. 1–95, 1989.

[JEN 76] JENSEN M., MECKLING W., "Theory of the firm: managerial behavior, agency costs and ownership structure", *Journal of Financial Economics*, vol. 3 no. 4, pp. 305–360, 1976.

[JEV 76] JEVONS W.S., *The Theory of Political Economy*, 2nd ed., Macmillan, London, 1976.

[JOH 87] JOHNSON G., *Strategic Change and the Management Process*, Blackwell, New York, 1987.

[JOH 92] JOHNSON H.T., *Relevance Regained: From Top-Down Control to Bottom-Up Empowerment*, Free Press, New York, 1992.

[KAG 83] KAGONO T., NONAKA I., SAKAKIBARA K. *et al.*, *Comparison of Management between Japan and USA*, Nihon Keizai Shinbun Inc., Tokyo, 1983.

[KAP 96] KAPLAN R.S., NORTON D.P., *The Balanced Scorecard. Translating Strategy Into Action*, Harvard Business School Press, Harvard, 1996.

[KAP 00] KAPLAN R.S., NORTON D.P., "Having trouble with your strategy? Then map it", *Harvard Business Review*, vol. 78, no. 5, pp. 167–176, 2000.

[KAP 01] KAPLAN R.S., NORTON D.P., *The Strategy-Focused Organization: How Balanced Scorecard Companies Thrive in the New Business Environment*, Harvard Business Press, Boston, 2001.

[KAR 15] KARABELL Z., "The uberization of money", *The Wall Street Journal*, available at: http://www.wsj.com/articles/the-uberization-of-finance-1446835102, 2015.

[KAW 12] KAWASAKI, T., "Current State and Issues of Accounting for SMEs (Small and Medium-sized Entities) in Japan", *Konan Kaikei Kenkyu*, vol. 6, pp. 1–10, 2012.

[KAW 14] KAWASAKI T., SAKAMOTO T., CHUBB O., *General Accounting Standard for Small- and Medium-Sized Entities in Japan*, Wiley Publishing Japan K.K., Tokyo, 2014.

[KAW 16] KAWASAKI T., *Accounting Theory for Small and Medium-sized Entities*, Cyuo Keizai Sha, Tokyo, 2016.

[KAY 15] KAYA D., KOCH M., "Countries' adoption of the International Financial Reporting Standard for Small and Medium-sized Entities (IFRS for SMEs) – early empirical evidence", *Accounting and Business Research*, vol. 45, no. 1, pp. 93–120, 2015.

[KAZ 89] KAZUSA Y., *American Management Accounting History (1)*, Doubunkan, Tokyo, 1989.

[KAZ 03] KAZUSA Y., "Payback period methods considering the cost of capital: discounted and premium payback period method", *Journal of Japanese Management Accounting*, vol. 12, no. 1, pp. 41–52, 2003.

[KIL 14] KILIÇ M., UYAR A., ATAMAN B., "Preparedness for and perception of IFRS for SMEs: evidence from Turkey", *Accounting and Management Information Systems*, vol. 13, no. 3, pp. 492–519, 2014.

[KIM 05] KIM J., MAHONEY J.T., "Property rights theory, transaction costs theory, and agency theory: an organizational economics approach to strategic management", *Managerial and Decision Economics*, vol. 26, no. 4, pp. 223–242, 2005.

[KIM 10] KIM J., MAHONEY J.T., "A strategic theory of the firm as a nexus of incomplete contracts: a property rights approach", *Journal of Management*, vol. 36, no. 4, pp. 806–832, 2010.

[KIN 10] KING N., HORROCKS C., *Interviews in Qualitative Research*, SAGE Publications, London, 2010.

[KOF 13] KOFAX, "Case study (Orange-France Telecom)", available at: http://www.kofax.com/~/media/Files/Kofax/Case-Studies/cs-orange-france-telecom-en.pdf, 2013.

[KRA 11] KRAAIJENBRINK J., SPENDER J.C., "Theories of the firm and their value creation assumptions", *SMS 31st Annual International Conference*, Miami, 2011.

[KUS 14] KUSHIBE S., "The movements of accounting standards for small and medium-sized enterprises in Japan", *Accounting*, (櫛部幸子「我が国における中小企業会計基準の動向」『會計』, vol. 185, no. 3, pp. 389–402, 2014.

[KUS 15] KUSHIBE S., "Usability and future applicability of the basic guidance of accounting for SMEs in Japan", *Journal of Accounting Research for Small- and Medium-Sized Entities (SMEs)*, vol. 1, no. 1, pp. 16–27, 2015.

[KYO 94] KYOCERA CORPORATION, Kyocera Philosophy Pocketbook, available at: global. kyocera.com, 1994.

[KYO 17] KYOCERA CORPORATION, *Inamori twelve management principles*, Inamori Kazuo Official Site (in Japanese), available at: http://www.kyocera.co.jp/inamori/management/twelve.html, 2017.

[LAM 98] LAMB M., NOBES C., ROBERTS A., "International variations in the connections between tax and financial reporting", *Accounting Business Research*, vol. 28, no. 3, pp. 173–188, 1998.

[LAN 97] LANGFIELD-SMITH K., "Management control systems and strategy: a critical review", *Accounting, Organizations and Society*, vol. 22, no. 2, pp. 207–232, 1997.

[LAR 14] LARDON, A., DELOOF M., "Financial disclosure by SMEs listed on a semi-regulated market: evidence from the Euronext Free Market", *Small Business Economics*, vol. 42, no. 2, pp. 361–385, 2014.

[LAT 87] LATOUR B., *Science in Action*, Harvard University Press, New York, 1987.

[LAT 88] LATOUR B., *The Pasteurization of France*, Harvard University Press, Cambridge, 1988.

[LAW 16] LAWSON R., "What do management accountants do?", *Institute of Management Accountants (IMA), Document 7-19-16,* available at: https://www.imanet.org/-/media/24e77e5955534d4c8e6d5ceec9fee 566.ashx?la=en, 2016.

[LEN 97] LENZ B., "Das Filière-Konzept als Analyseinstrument der organisatorischen und räumlichen Anordnung von Produktions- und Distributionsprozessen", *Geographische Zeitschrift*, vol. 85, no. 1, pp. 20–33, 1997.

[LEU 00] LEUZ C., VERRECCHIA R., "The economic consequences of increased disclosure", *Journal of Accounting Research*, vol. 38, pp. 91–124, 2000.

[LIB 10] LIBBY T., LINDSAY R.M., "Beyond budgeting or budgeting reconsidered? A survey of North-American budgeting practice", *Management Accounting Research*, vol. 21, no. 1, pp. 56–75, 2010.

[LIN 00] LINDER J., CANTRELL S., Changing Business Models: Surveying the Landscape, Working paper, Accenture Institute for Strategic Change, 2000.

[LIT 12] LITIJENS R., BISSESSUR S., LANGENDIJK H. *et al.*, "How do preparers perceive costs and benefits of IFRS for SMEs? Empirical evidence from the Netherlands", *Accounting in Europe*, vol. 9, no. 2, pp. 227–250, 2012.

[LOE 15] LOEBBECKE C., PICOT A., "Reflections on societal and business model transformation arising from digitization and big data analytics: a research agenda", *The Journal of Strategic Information Systems*, vol. 24, no. 3, pp. 149–157, 2015.

[LUS 06] LUSCH R.F., VARGO S.L., "Service-dominant logic: reactions, reflections and refinements", *Marketing Theory*, vol. 6, no. 3, pp. 281–288, 2006.

[LYN 03] LYNCH R.F., *Corporate Strategy*, Prentice Hall Financial Times, Harlow, 2003.

[MAC 00] MACMILLAN H., TAMPOE M., *Strategic Management*, Oxford University Press, Oxford, 2000.

[MAG 02] MAGRETTA J. "Why business models matter", *Harvard Business Review*, vol. 80, no. 5, pp. 86–92, 2002.

[MAL 08] MALMI T., BROWN D.A., "Management control systems as a package: opportunities, challenges and research directions", *Management Accounting Research*, vol. 19, no. 4, pp. 287–300, 2008.

[MAN 12] MANDAI K., "The Necessity of Coexistence for Both ASBJ Guidelines and General standards", *Corporate Accounting*, vol. 64, no. 4, pp. 32–39, 2012.

[MAR 02] MARGINSON D.E.W., "Management control systems and their effects on strategy formation at middle-management levels: evidence from a UK organization", *Strategic Management Journal*, vol. 23, no. 11, pp. 1019–1031, 2002.

[MAR 98] MARSHALL A., *Elements of Economics of Industry*, MacMillan, London, 1898.

[MAR 91] MARSTON C.L., SHRIVES P.J., "The use of disclosure indices in accounting research: a review article", *British Accounting Review*, vol. 23, no. 3, pp. 195–210, 1991.

[MAT 05] MATSUGI S., "Analysis of management control in micro-profit centre system", *Cost Accounting Research*, vol. 29, no. 1, pp. 92–104, 2005.

[MCK 10] MCKRAW T., *Prophet of Innovation: Joseph Schumpeter and Creative Destruction*, Belknap Press, Cambridge, 2010.

[MEN 00] MENDELSON H., "Organizational architecture and success in the information technology industry", *Management Science*, vol. 46, no. 4, pp. 513–529, 2000.

[MEN 71] MENGER C., *Grundsätze der Volkswirthschaftslsehre*, Braumüller, Vienna, (published 1871) (English translation: New York University Press, New York, 1976).

[MER 98] MERCHANT K.A., *Modern Management Control Systems: Text and Cases*, Prentice Hall, Upper Saddle River, 1998.

[MIN 98] MINTZBERG H., AHLSTRAND B., LAMPEL J., *Strategy Safari: A Guided Tour through the Wilds of Strategic Management*, Free Press, New York, 1998.

[MIS 14] MISTRY V., SHARMA U., LOW M., "Management accountants' perception of their role in accounting for sustainable development : An exploratory study", *Pacific Accounting Review*, vol. 26, nos. 1–2, pp. 112–133, 2014.

[MIY 03] MIYA H., *Amoeba Management Theory: The Mechanism and Introduction of Micro-Profit Center*, Toyo Keizai Shinposha, Tokyo (in Japanese), 2003.

[MON 11] MONGAN K., "*Cloud Computing – The Storm is Coming*", *Accountancy Ireland*, vol. 43, no. 3, pp. 58–59, 2011.

[MOR 88] MORGAN G., "Accounting as reality construction: towards a new epistemology for accounting", *Accounting, Organizations and Society*, vol. 13, no. 5, pp. 477–486, 1988.

[MOR 87] MORRIS R.D., "Signalling, agency theory and accounting policy choice", *Accounting and Business Research*, vol. 18 no. winter, pp. 47–56, 1987.

[MOR 05] MORRIS M., SCHINDEHUTTE M., ALLEN J., "The entrepreneur's business model: Toward a unified perspective", *Journal of Business Research*, vol. 58, no. 6, pp. 726–735, 2005.

[MOR 85] MORVAN Y., *Fondements d'Economie industrielle*, Economica, Paris, 1985.

[MUL 10] MULLEROVA L., PASEKOVÁ M., KUBICKOVA D., Analysis of differences in reporting according to IFRS in SMEs in the Czech Republic and its influence on performance measurement, Working Paper, available at SSRN: https://ssrn.com/abstract=2221633, 2010.

[MUR 10] MURATA N., "Empirical research for practices in small and middle-sized entities accounting", *Sangyo Keiei Project Houkoku (Nihon University)*, vol. 33, no. 2, pp. 1–165, 2010.

[NOB 10] NOBES C., "On Researching into the Use of IFRS by Private Entities in Europe," *Accounting in Europe*, vol. 7, no. 2, pp. 213–226, 2010.

[NON 07] NONAKA I., TOYAMA R., "Strategic management as distributed practical wisdom (phronesis)", *Industrial and Corporate Change*, vol. 16, no. 3, pp. 371–394, 2007.

[NON 14] NONAKA I., KODAMA M., HIROSE A. *et al.*, "Dynamic fractal organizations for promoting knowledge-based transformation: a new paradigm for organizational theory", *European Management Journal*, vol. 32, no. 1, pp. 137–146, 2014.

[NON 95] NONAKA I., TAKEUCHI H., *The Knowledge-Creating Company: How Japanese Companies Create the Dynamics of Innovation*, Oxford University Press, New York, 1995.

[OEC 17] OECD, Key issues for digital transformation in the G20, Report prepared for a joint G20 German Presidency/OECD Conference Berlin, Germany, available at: https://www.oecd.org/g20/key-issues-for-digital-transformation-in-the-g20.pdf, 2017.

[OKA 07] OKANO H., SUZUKI T., "A history of Japanese management accounting", in CHAPMAN C.S., HOPWOOD A.G., SHIELDS M.D. (eds), *Handbook of Management Accounting Research*, Oxford University Press, New York, 2007.

[OST 05] OSTERWALDER A., PIGNEUR Y., Tucci C. L., "Clarifying Business Models: Origins, Present, and Future of the Concept", *Communications of the Association for Information Systems*, vol. 16, no. 1, pp. 1–25, 2005.

[OST 10] OSTERWALDER A., PIGNEUR Y., *Business Model Generation: A Handbook for Visionaries, Game Changers, and Challengers*, John Wiley & Sons, New York, 2010.

[OTL 99] OTLEY D., "Performance management: a framework for management control systems research", *Management Accounting Research*, vol. 10, no. 4, pp. 363–382, 1999.

[OUC 79] OUCHI W.G., "A conceptual framework for the design of organizational control mechanisms", *Management Science*, vol. 25, no. 9, pp. 833–848, 1979.

[PAT 00] PATEL K., MCCARTHY M.P., *Digital Transformation: The Essentials of E-Business Leadership,* McGraw-Hill, London, 2000.

[PAT 07] PATELLI P., PRENCIPE A., "The Relationship between Voluntary Disclosure and Independent Directors in the Presence of a Dominant Shareholder", *European Accounting Review*, vol. 16 no. 1, pp. 5-33, 2007.

[PAT 15] PATIL H., "FOUR signs the uberization of accounting has already begun", available at: http://cpatrendlines.com/2015/06/15/patil-uberization-accounting/, 2015.

[PER 09] PERRINI F., RUSSO A., TENCATI A. *et al.*, Going beyond a long-lasting debate: what is behind the relationship between corporate social and financial performance?, Working paper, Cranfield School of Management SDA Boccori, Vlerick Leuven Gent Business School, Milan, 2009.

[PET 01] PETEROVIC O., KITTL C., TEKSTEN R.D., "Developing Business Models for eBusiness", *International Conference on Electronic Commerce*, Vienna, 2001.

[POL 66] POLANYI M., *The Tacit Dimensions*, Routledge and Kegan Paul, London, 1966.

[POR 79] PORTER M.E., "How competitive forces shape strategy", *Harvard Business Review*, vol. 57, no. 2, pp. 137–145, 1979.

[POR 80] PORTER M.E., *Competitive Strategy: Techniques for Analyzing Industries and Competitors*, Free Press, New York, 1980.

[POR 81] PORTER M.E., "The contributions of industrial organization to strategic management", *The Academy of Management Review*, vol. 6, no. 4, pp. 609–620, 1981.

[POR 85a] PORTER M.E., *Competitive Strategy: Creating and Sustaining Superior Performance*, Free Press, New York, 1985.

[POR 85b] PORTER M.E., *Competitive Advantage*, Free Press, New York, 1985.

[POR 90] PORTER M.E., *The Competitive Advantage of Nations*, Free Press, New York, 1990.

[POR 91] PORTER M.E., "Towards a dynamic theory of strategy", *Strategic Management Journal*, vol. 12, pp. 95–117, 1991.

[POR 96] PORTER M.E., "What is strategy?", *Harvard Business Review*, vol. 74, no. 6, pp. 61–78, 1996.

[POR 01] PORTER M.E., "Strategy and the Internet", *Harvard Business Review*, vol. 79, no. 3, pp. 63–78, 2001.

[POR 11] PORTER M.E., KRAMER M.R., "Creating shared value: how to reinvent capitalism – and unleash a wave of innovation and growth – and unleash a wave of innovation and growth", *Harvard Business Review*, vol. 89, no. 1, pp. 1–17, 2011.

[PRE 04] PRENCIPE A., "Proprietary Costs and Determinants of Voluntary Segment Disclosure: Evidence from Italian Listed Companies", *European Accounting Review*, vol. 13, no. 2, pp. 319–340, 2004.

[PRI 13] PRIVATE COMPANY COUNCIL (PCC), Private Company Decision-Making Framework: A Guide for Evaluating Financial Accounting and Reporting for Private Companies, Financial Accounting Standard Board, New York, 23 December, 2013.

[QUA 12] QUAGLI A., PAOLONI P., "How is the IFRS for SME accepted in the European context? An analysis of the homogeneity among European countries, users and preparers in the European commission questionnaire", *Advances in Accounting, Incorporating Advances in International Accounting*, vol. 28, no. 1, pp. 147–156, 2012.

[QUI 80] QUINN J.B., *Strategies for Change: Logical Incrementalism*, Irwin Professional Publishing, Chicago, 1980.

[QUI 14] QUINN M., STRAUB E., KRISTANDL G., "The effects of cloud technology on management accounting and decision-making", *Financial Management*, vol. 10, no. 6, pp. 54–55, 2014.

[RAI 00] RAIKES P., JENSEN M.F, PONTE S., Global commodity chain analysis and the French Filière approach: comparison and critique, Working Paper, Centre for Development Research (CDR), Copenhagen, 2000.

[RAP 01] RAPPA M., *Managing the digital enterprise – Business models on the Web*, North Carolina State University, Raleigh, 2001.

[REM 93] REMER D.S., STOKDYK S.B., VAN DRIEL M., "Survey of project evaluation techniques currently used in industry", *International Journal of Production Economics*, vol. 32, pp. 103–115, 1993.

[RIC 51] RICARDO D., *Principles of Political Economy and Taxation*. (published 1817), in SRAFFA P., DOBB M. (eds), *The Works and Correspondence of David Ricardo*, Cambridge University Press, Cambridge, 1951.

[ROB 96] ROBSON K., "Accounting numbers as 'inscription': action at a distance and the development of accounting", *Accounting, Organizations and Society*, vol. 17, no. 7, pp. 685–708, 1996.

[ROB 01] ROBSON P.J.A., BENNETT R.J., "SME growth: the relationship with business advice and external collaboration", *Small Business Economics*, vol. 15, no. 3, pp. 193–208, 2001.

[ROM 86] ROMNEY A., BATCHELDER W., WELLER S., "Culture as consensus: a theory of culture and informant accuracy", *Americana Anthropologist*, vol. 88, no. 2, pp. 313–338, 1986.

[RYA 02] RYAN P.A., RYAN G.P., "Capital budgeting practices of the fortune 1000: how have things changed?", *Journal of Business and Management*, vol. 8, no. 4, pp. 355–364, 2002.

[SAW 08] SAWABE N., TOBITA T., "A field survey on management philosophy, social relation, management accounting and corporate performance", *Accounting*, vol. 60, no. 12, pp. 133–141 (in Japanese), 2008.

[SCH 01] SCHATZKI T., KNORR CETINA K., SAVIGNY E., *The Practice Turn in Contemporary Theory*, Routledge, London, 2001.

[SCH 34] SCHUMPETER J., *The Theory of Economic Development*, Harvard University Press, Cambridge, 1934.

[SCH 42] SCHUMPETER J., *Capitalism, Socialism and Democracy*, Harper, New York, 1942.

[SER 07] SERRANO R., Cooperative games: core and Shapley value, CEMFI Working Paper No. 0709, available at: http://www.cemfi.es/ftp/wp/0709.pdf, 2007.

[SHA 92] SHANK J.K., GOVINDARAJAN V., "Strategic cost management: the value chain perspective", *Journal of Management Accounting Research*, vol. 4, no. 4, pp. 179–197, 1992.

[SHA 93] SHANK J.K., GOVINDARAJAN V., *Strategic Cost Management: The New Tool For Competitive Advantage*, Free Press, New York, 1993.

[SHA 53] SHAPLEY L.S., "A value for n-person games," in KUHN H.W., TUCKER A.W. (eds), *Contributions to the Theory of Games II (Annals of Mathematics Studies 28)*, Princeton University Press, Princeton, 1953.

[SHI 07] SHIMIZU N., KATO Y., SAKAGUCHI J. *et al.*, "Field survey on capital investment management (1)", *Kigyokaikei*, vol. 59, no. 8, pp. 105–111, 2007.

[SIM 90] SIMONS R., "The role of management control systems in creating competitive advantage: new perspectives", *Accounting, Organizations and Society*, vol. 15, nos. 1–2, pp. 127–143, 1990.

[SIM 95] SIMONS R., *Levers of control: How Managers Use Innovative Control Systems to Drive Strategic Renewal*, Harvard Business Press, Boston, 1995.

[SIM 00] SIMONS R., *Performance Measurement & Control Systems for Implementing Strategy: Text & Cases*, Prentice Hall, Upper Saddle River, 2000.

[SMI 81] SMITH A., "An inquiry into the nature and causes of the wealth of nations", (published 1776), in CAMBELL R.H., SKINNER A.J. (eds), Liberty Fund, Indianapolis, 1981.

[SPE 73] SPENCE M., "Job market signalling", *Quarterly Journal of Economics*, vol. 87, no. 3, pp. 355–374, 1973.

[STE 10] STEENKAMP N., KASHYAP C., "Importance and contribution of intangible assets: SME managers' perceptions", *Journal of Intellectuel Capital*, vol. 11, no. 3, pp. 368–390, 2010.

[STI 66] STIGLER G., *The Theory of Price*, 3rd ed., Macmillan, New York, 1966.

[STI 89] STIGLER G., "Two notes on the Coase theorem", *Yale Law Journal*, vol. 99, no. 3, pp. 631–633, 1989.

[STO 04] STOLTERMAN E., FOREST A.C., "Information technology and the good life", in KAPLAN B., TRUEX D.P., WASTELL D., *et al.* (eds), *Information Systems Research: Relevant Theory and Informal Practice*, Springer, Berlin, 2004.

[SUN 13] SUNDERS J.C., LINDBERG D.L., SEIFERT D.L., "Is IFRS for SMEs a Beneficial Alternative for Private Companies?," *The CPA Journal,* no. Feb., pp. 32–35, 2013.

[TAN 99] TANI T., "Empowerment by micro-profit center: the case of amoeba management", *Kokumin-Keizai-Zasshi*, vol. 180, no. 5, pp. 47–59, 1999.

[TAP 00] TAPSCOTT D., TICOLL D., LOWY A., *Digital Capital – Harnessing the Power of Business Webs*, Harvard Business School Press, Boston, 2000.

[TEE 10] TEECE D., "Business models, business strategy and innovation", *Long Range Planning*, vol. 43, no. 2/3, pp. 172–194, 2010.

[TEI 13] TEIKOKU DATA BANK, A Report for Empirical Investigation for Small and Medium-size Entities Accounting in 2012, Minister of Economy, Trade and Industry (METI), available at: http://www.meti.go.jp/meti_lib/report/2013fy/E003483.pdf, 2013.

[TIM 98] TIMMERS P., "Business Models for Electronic Markets", *Journal on Electronic Markets*, vol. 8, no. 2, pp. 3–8, 1998.

[TOR 16] TORCHIA M., CALABRO A., "Board of directors and financial transparency and disclosre: Evidence from Italy", *Corporate Governance*, vol. 16, no. 3, pp. 593–608, 2016.

[UEN 12a] UENISHI S., "The role of tax accountant", in KAWASAKI T., MANDAI K. (eds), *The General Standard for Small and Middle-sized Entities*, Cyuo Keizai Sha, Tokyo, 2012.

[UEN 12b] UENISHI S., "The impact of setting the general standards for SMEs accounting upon SMEs practices", *Zeikei*, vol. 28, no. 1, pp. 42–49, 2012.

[UEZ 14] UEZU Y., "A study on accounting standards of small and medium enterprises", *Faculty of Law and Letters (University of the Ryukyus)*, vol. 87, pp. 33–43, 2014.

[UYA 13] UYAR A., GÜNGÖRMÜS A.H., "Receptions and knowledge of accounting professionals on IFRS for SMEs: Evidence from Turkey", *Research in Accounting Regulation*, vol. 25, no. 1, pp. 77–87, 2013.

[VAN 09] VAN DER LANN S., "The role of theory in explaining motivation for corporate social disclosures: voluntary disclosures vs solicited disclosures", *Australiasian Accounting Business and Finance Journal*, vol. 3, no. 4, pp. 15–29, 2009.

[VAR 04] VARGO R.F., LUSCH S.L., "Evolving for a new dominant logic for marketing", *Journal of Marketing*, vol. 68, pp. 1–17, 2004.

[VAR 08] VARGO R.F., LUSCH S.L., "From goods to service(s): divergences and convergences of logics", *Industrial Marketing Management*, vol. 37, no. 3, pp. 254–259, 2008.

[VEN 98] VENKATRAMAN N., HENDERSON J.C., "Real strategies for virtual organizing", *Sloan Management Review*, vol. 40, no. 1, pp. 33–48, 1998.

[VER 83] VERRECCHIA R., "Discretionary disclosure", *Journal of Accounting and Economics*, vol. 5, pp. 179–194, 1983.

[WAL 74a] WALRAS L., "Principe d'une théorie mathématique de l'échange", *Séances et travaux de l'Académie des Sciences morales et politiques*, vol. 101, no. 33, *Journal des Économistes*, vol. 100, pp. 5–31, 1874.

[WAL 74b] WALRAS L., *Éléments d'économie politique pure*, Corbaz, Lausanne, 1874.

[WAT 04] WATANABE T., "The effect of line-company system for spontaneous motivation of company-leaders: positive research of line-company system in Sumitomo Electronic Inc.", *Cost Accounting Research*, vol. 28, no. 2, pp. 12–26, 2004.

[WAT 86] WATTS R.L., ZIMMERMAN J.L., *Positive Accounting Theory*, Englewood Cliffs, Prentice-Hall, Upper Saddle River, 1986.

[WEH 14] WEHRFRITZ M., HALLER A., "National influence on the application of IFRS: Interpretations and accounting estimates by German and British accounts", *Advances in Accounting, incorporating Advances in International Accounting*, vol. 30, no. 1, pp. 196–208, 2014.

[WEI 01] WEILL P., VITALE M., *Place to space: Migrating to eBusiness Models*, Harvard Business School Press, Boston, 2001.

[WEI 69] WEINGARTNER H.M., "Some new views of the payback period and capital budgeting decisions", *Management Science*, vol. 15, no. 12, pp. 594–607, 1969.

[WES 11] WESTERMAN G., CALMÉJANE L., BONNET D., *et al.*, "Digital Transformation: A Roadmap for Billion-Dollar Organizations", available at: https://www.capgemini.com/resource-file-access/resource/pdf/Digital_Transformation_A_Roadmap_for_Billion_Dollar_Organizations.pdf, 2011.

[WIL 81] WILLIAMSON O.E., "The economics of organization: the transaction cost approach", *American Journal of Sociology*, vol. 87, no. 3, pp. 548–577, 1981.

[WHI 96] WHITTINGTON R., "Strategy as practice", *Long Range Planning*, vol. 29, no. 5, pp. 731–735, 1996.

[YAN 12] YANAGA M., "Small and middle-sized entities: accounting from legal perspective", in KAWASAKI T., MANDAI K. (eds), *The General Standard for Small and Middle-sized Entities*, Cyuo Keizai Sha, Tokyo, 2012.

[YIP 04] YIP G., "Using strategy to change your business model", *Business Strategy Review*, vol. 15, no. 2, pp. 17–24, 2004.

[YU 05] YU C.-C., "Linking the balanced scorecard to business models for value-based strategic management in e-business", in BAUKNECHT K., BICHLER M., PROLL B. (eds), *EC-Web 2005*, Springer-Verlag, Berlin, 2005.

Index

A

B, C

D, E, F

G, H

I

J, K, L

M, O, P

R, S

T

U, V

Other titles from

iSTE

in

Innovation, Entrepreneurship and Management

2017

CHALENÇON Ludivine
Location Strategies and Value Creation of International Mergers and Acquisitions

CHAUVEL Danièle, BORZILLO Stefano
The Innovative Company: An Ill-defined Object
(Innovation Between Risk and Reward Set – Volume 1)

CORSI Patrick
Going Past Limits To Growth

D'ANDRIA Aude, GABARRET Inés
Building 21st Century Entrepreneurship
(Innovation and Technology Set – Volume 2)

LEBERT Didier, EL YOUNSI Hafida
International Specialization Dynamics
(Smart Innovation Set – Volume 9)

MAESSCHALCK Marc
Reflexive Governance for Research and Innovative Knowledge
(Responsible Research and Innovation Set – Volume 6)

MASSOTTE Pierre
Ethics in Social Networking and Business 1

MASSOTTE Pierre, CORSI Patrick
Smart Decisions in Complex Systems

MEDINA Mercedes, HERRERO Mónica, URGELLÉS Alicia
Current and Emerging Issues in the Audiovisual Industry
(Diverse and Global Perspectives on Value Creation Set – Volume 1)

MICHAUD Thomas
Innovation, Between Science and Science Fiction
(Smart Innovation Set – Volume 10)

SAVIGNAC Emmanuelle
The Gamification of Work: The Use of Games in the Workplace

UZUNIDIS Dimitri, SAULAIS Pierre
Innovation Engines: Entrepreneurs and Enterprises in a Turbulent World
(Innovation in Engineering and Technology Set – Volume 1)

2016

BARBAROUX Pierre, ATTOUR Amel, SCHENK Eric
Knowledge Management and Innovation
(Smart Innovation Set – Volume 6)

BEN BOUHENI Faten, AMMI Chantal, LEVY Aldo
Banking Governance, Performance And Risk-Taking:
Conventional Banks Vs Islamic Banks

BOUTILLIER Sophie, CARRÉ Denis, LEVRATTO Nadine
Entrepreneurial Ecosystems
(Smart Innovation Set – Volume 2)

BOUTILLIER Sophie, UZUNIDIS Dimitri
The Entrepreneur
(Smart Innovation Set – Volume 8)

BOUVARD Patricia, SUZANNE Hervé
Collective Intelligence Development in Business

GALLAUD Delphine, LAPERCHE Blandine
Circular Economy, Industrial Ecology and Short Supply Chains
(Smart Innovation Set – Volume 4)

GUERRIER Claudine
Security and Privacy in the Digital Era
(Innovation and Technology Set – Volume 1)

MEGHOUAR Hicham
Corporate Takeover Targets

MONINO Jean-Louis, SEDKAOUI Soraya
Big Data, Open Data and Data Development
(Smart Innovation Set – Volume 3)

MOREL Laure, LE ROUX Serge
Fab Labs: Innovative User
(Smart Innovation Set – Volume 5)

PICARD Fabienne, TANGUY Corinne
Innovations and Techno-ecological Transition
(Smart Innovation Set – Volume 7)

2015

CASADELLA Vanessa, LIU Zeting, UZUNIDIS Dimitri
Innovation Capabilities and Economic Development in Open Economies
(Smart Innovation Set – Volume 1)

CORSI Patrick, MORIN Dominique
Sequencing Apple's DNA

CORSI Patrick, NEAU Erwan
Innovation Capability Maturity Model

FAIVRE-TAVIGNOT Bénédicte
Social Business and Base of the Pyramid

GODÉ Cécile
Team Coordination in Extreme Environments

MAILLARD Pierre
Competitive Quality and Innovation

MASSOTTE Pierre, CORSI Patrick
Operationalizing Sustainability

MASSOTTE Pierre, CORSI Patrick
Sustainability Calling

2014

DUBÉ Jean, LEGROS Diègo
Spatial Econometrics Using Microdata

LESCA Humbert, LESCA Nicolas
Strategic Decisions and Weak Signals

2013

HABART-CORLOSQUET Marine, JANSSEN Jacques, MANCA Raimondo
VaR Methodology for Non-Gaussian Finance

2012

DAL PONT Jean-Pierre
Process Engineering and Industrial Management

MAILLARD Pierre
Competitive Quality Strategies

POMEROL Jean-Charles
Decision-Making and Action

SZYLAR Christian
UCITS Handbook

2011

LESCA Nicolas
Environmental Scanning and Sustainable Development

LESCA Nicolas, LESCA Humbert
Weak Signals for Strategic Intelligence: Anticipation Tool for Managers

MERCIER-LAURENT Eunika
Innovation Ecosystems

2010

SZYLAR Christian
Risk Management under UCITS III/IV

2009

COHEN Corine
Business Intelligence

ZANINETTI Jean-Marc
Sustainable Development in the USA

2008

CORSi Patrick, DULIEU Mike
The Marketing of Technology Intensive Products and Services

DZEVER Sam, JAUSSAUD Jacques, ANDREOSSO Bernadette
Evolving Corporate Structures and Cultures in Asia / Impact of Globalization

2007

AMMI Chantal
Global Consumer Behavior

2006

BOUGHZALA Imed, ERMINE Jean-Louis
Trends in Enterprise Knowledge Management

CORSI Patrick *et al.*
Innovation Engineering: the Power of Intangible Networks

Printed and bound by CPI Group (UK) Ltd, Croydon, CR0 4YY
13/09/2026
14961055-0001